THEORY OF
ORBITAL
MOTION

THEORY OF
ORBITAL
MOTION

ARJUN TAN

Alabama A & M University, USA

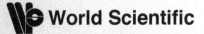

World Scientific

NEW JERSEY · LONDON · SINGAPORE · BEIJING · SHANGHAI · HONG KONG · TAIPEI · CHENNAI

Published by

World Scientific Publishing Co. Pte. Ltd.

5 Toh Tuck Link, Singapore 596224

USA office: 27 Warren Street, Suite 401-402, Hackensack, NJ 07601

UK office: 57 Shelton Street, Covent Garden, London WC2H 9HE

British Library Cataloguing-in-Publication Data
A catalogue record for this book is available from the British Library.

Illustrations by Sonya Lyatskaya.

THEORY OF ORBITAL MOTION

ISBN-13 978-981-270-911-0
ISBN-10 981-270-911-8
ISBN-13 978-981-270-912-7
ISBN-10 981-270-912-6 (pbk)

Typeset by Stallion Press
Email: enquiries@stallionpress.com

Printed in Singapore.

Dedicated to the Memory of Mother, Father and Sister

Preface

Orbital motion commonly means the motion of an object around a second one which is infinitely more massive than the first. But it can also mean the motion of two objects of comparable size around their center of mass. Without this basic motion, the solar system as we know would not exist and life on the planet Earth would not be possible. It is this elemental phenomenon which was largely responsible for Newton's discovery of his law of gravitation. It is also the same phenomenon which necessitated his invention of the Calculus, which continues to be the backbone of Mathematics today.

Orbital motion is mainly taught as a small portion of a Mechanics course in colleges and universities in the traditional Physics curriculum. Hence there is a real scarcity of books devoted to this matter. There exist advanced books on this subject for the specialists engaged in Celestial Mechanics or spaceflight, which are seldom used as textbooks in colleges or universities. This void was even felt in the United States Air Force Academy, where one faculty came up with the solution by publishing their own textbook [*Fundamentals of Astrodynamics* by Bate *et al.* (1971)].

The origin of this book dates back to the year 2001, when the author initiated the Space Science Program at Alabama A & M

University, leading to the B.S. and M.S. degrees in Physics with Space Science as a specialization area. This effort was supported by two successive NASA Grants, jointly funded by the Office of Space Science and the Office of Equal Opportunity. One of the courses in this program was *Introductory Orbital Mechanics*, which is an upper undergraduate level course. The need for a proper textbook was immediate. There was only one lower level book *Elements of Astromechanics* by Van de Kamp (1964), which was deemed inadequate. Two other books *Fundamentals of Celestial Mechanics* by Danby (1988) and *Fundamentals of Astrodynamics* by Bate *et al.* (1971) were graduate level textbooks intended for Aerospace Engineering students. Thus comes the conception of this book.

This book is theoretically oriented. Most of the results are derived in the book from the first principles. Unnecessary lengthy discussions on physical principles and concepts are not given space in this book. References are cited wherever applicable. Nonetheless, the need to look for references elsewhere is kept to a minimum. It is written as an all-in-one book for the upper undergraduate students in Physics. The author entertains the hope that it will serve as a user-friendly book for the student and a useful reference book for the specialist.

The author wishes to express his gratitude to the Alabama A & M Univesity for granting him a Sabbatical Leave, during which much of the book was completed.

Arjun Tan
Department of Physics
Alabama A & M University
Box 447
Normal, Alabama 35762
U.S.A.

Contents

Kepler's Laws of Planetary Motion

1.1 Background

Man has gazed at the night sky from time immemorial. What was perhaps the most conspicuous to early man was a star field which revolved around the pole star but appeared to have a fixed configuration. Against this background, the Sun and the Moon, of course, rose and set at different locations. But more interesting was the inexplicable back-and-forth motion of the planets relative to the star field, which earned them their name "the wanderers". The subject of planetary motion attracted astronomers and mathematicians alike. It was Tycho Brahe who compiled an extensive catalog of the locations of the planets before the advent of the telescope. And it was from this catalog that his pupil Johannes Kepler was to eventually deduce his laws governing the motion of planets around the Sun.

1.2 Kepler's Laws of Planetary Motion

Kepler enunciated his three laws of planetary motion based on deductions from astronomical data on Mars taken by Tycho Brahe. The first two laws were published in 1609, and the third law ten years later in 1619. The following are unambiguous statements of Kepler's laws:

Law 1. The orbit of a planet around the Sun is an ellipse, with the Sun being situated at one focus of that ellipse.

Law 2. The radius vector of the planet from the Sun sweeps out equal areas in equal times, i.e., the 'areal velocity' of the planet is a constant. This is called the *law of areas*.

Law 3. The square of the period of a planet is directly proportional to the cube of the semi major axis. This is known as the *harmonic law*. Rendered symbolically:

$$P^2 \propto a^3, \qquad\qquad (1.1)$$

where the symbols carry their obvious meanings.

The first two laws are concerned with individual orbits. The first law indicates the shape of the orbit and the location of the Sun within that orbit. It does not contain time. The second law indicates the speed with which the planet traverses the orbit at different locations. The third law enables comparisons between different orbits.

Kepler's laws are independent laws of nature. The second law does is not contained in the first and does not follow from the first as a corollary. Likewise, the third law provides information not contained in the first two. However, all three of Kepler's laws can be derived from Newton's law of gravitation, which therefore, is a more fundamental law of nature. With the sole exception of Archimedes principle, Kepler's laws are the oldest laws found in physics textbooks today.

1.3 Keplerian Ellipse

It is easy to visualize the relevant geometrical quantities associated with the elliptical orbit of a planet.

In Fig. 1.1, *PLBMAM'CL'* is the orbital ellipse of the planet. An *ellipse* has two *foci* (plural of *focus*). The Sun is located at one focus *F*. The conjugate focus *E* is unoccupied (*empty focus*). *O* is the center of the ellipse. A diameter is any chord passing through the center *O*. The longest diameter *AOP* is called the *major axis* and is denoted by 2*a*, so that half of the major axis (the *semi major axis*) is equal to *a*. Likewise, the shortest diameter *BOC* is called the *minor axis* and is denoted by 2*b*, such that half of the minor axis (the *semi minor axis*) is equal to *b*. Two other chords of interest *LFL'* and *MEM'* are those passing through the foci and perpendicular to the major axis (and thus parallel with the minor axis). They are called the *latera recta* (plural of *latus rectum*) and denoted by 2*p*, such that half of a latus rectum (the *semi latus rectum*) is *p*.

In our solar system, the planets revolve around the Sun in the *counter-clockwise* direction when viewed from the northern sky.

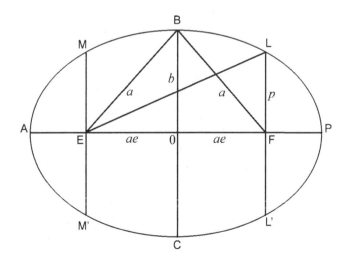

Fig. 1.1

Curiously, this fits into our convention of a **right-handed system.** The **perihelion** P marks the point of nearest approach of the planet to the Sun. The **aphelion** A is the point of farthest retreat of the planet from the Sun. The perihelion and aphelion are also termed **apsides** and the line joining them (the major axis) the **line of apsides** or **apsidal line.** By virtue of a unique property of the ellipse, the sum of the distances of any point on the ellipse from the two foci is a constant, which is equal to the major axis $2a$. This property is the basis of the **string and pins construction** of the ellipse (see Appendix A.5).

The shape of an ellipse (i.e., its **oblateness**) is given by its **eccentricity** which is defined as the ratio of the distance between the focal points to its major axis:

$$e = \frac{\overline{EF}}{\overline{AP}}. \tag{1.2}$$

The value of eccentricity of an ellipse lies in the range $0 \leq e < 1$. As e gets smaller, the ellipse becomes rounder and the two foci approach one another. When $e = 0$, the two foci coincide, and we get a circle. The circle is the limiting case of an ellipse whose eccentricity is 0. It is thus included in the family of ellipses. Kepler's first law permits perfectly circular orbits. On the other end of the scale, $e = 1$ represents the parabola, which is an open curve, and is therefore excluded from the family of ellipses.

An ellipse is completely defined by two parameters. They can be (a, b), (a, e), (p, e) or some other combination. The values for some relevant distances can be readily obtained from Euclidean geometry. From Eq. (1.2), we have $\overline{EO} = ae = \overline{OF}$. Also, $\overline{EB} = a = \overline{BF}$ and $\overline{EL} = 2a - p$. The relation between the semi minor axis b and the semi major axis a and that between the semi latus rectum p and the semi major axis a can be obtained from the Pythagoras' theorem applied to triangles BOF and ELF, respectively, and a special property of the ellipse which states that the sum of the focal distances of any point on the ellipse is a constant equal to the major axis $2a$ (see Appendix A.3):

$$b = a\sqrt{1 - e^2}, \tag{1.3}$$

and

$$p = a(1 - e^2). \tag{1.4}$$

1.4 Polar Equation of a Conic

Polar coordinates are the most natural and convenient coordinate system for planetary motion. The general equation of a conic in polar coordinates (r, θ) is given by

$$r = \frac{p}{1 + e\cos\theta}. \tag{1.5}$$

The great advantage of this equation is that it fits all conics with suitable choices of the eccentricity. Whereas planetary orbits are given by ellipses (which include the circle), escape trajectories are given by parabolic orbits and planetary fly-by trajectories are described by hyperbolic orbits.

Figure 1.2 shows the planetary orbit with the polar coordinates marked. Putting $\theta = 0, \pi, \pi/2$ and $3\pi/2$ in Eq. (1.5), one obtains the radial distances at the perihelion, aphelion and the ends of the latus

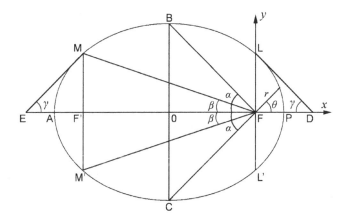

Fig. 1.2

rectum through the Sun:

$$r_P = a(1 - e), \tag{1.6}$$

$$r_A = a(1 + e), \tag{1.7}$$

and

$$r_L = r_{L'} = a(1 - e^2). \tag{1.8}$$

The radial distances at the ends of the minor axis and the latus rectum through the empty focus can be obtained from considerations of Sec. 1.2:

$$r_B = r_C = a, \tag{1.9}$$

and

$$r_M = r_{M'} = a(1 + e^2). \tag{1.10}$$

The angles α and β in Fig. 1.2 can be found in terms of e:

$$\alpha = \sin^{-1} \sqrt{1 - e^2} = \cos^{-1} e = \tan^{-1} \frac{\sqrt{1 - e^2}}{e}, \tag{1.11}$$

and

$$\beta = \sin^{-1} \frac{1 - e^2}{1 + e^2} = \cos^{-1} \frac{2e}{1 + e^2} = \tan^{-1} \frac{1 - e^2}{2e}. \tag{1.12}$$

The angular coordinates of B, M, M' and C are thus $\pi - \alpha$, $\pi - \beta$, $\pi + \beta$ and $\pi - \beta$, respectively.

1.5 The Slope of a Tangent to an Ellipse

The velocity of a planet is necessarily tangential to its orbital ellipse. In calculating the velocities at specific points, one often needs the slope of the tangent at those points. This is most conveniently done in Cartesian coordinates (x, y), in which the slope is simply dy/dx.

The transformation from polar to Cartesian coordinates is facilitated by the relations

$$x = r\cos\theta = \frac{p\cos\theta}{1 + e\cos\theta}, \qquad (1.13)$$

and

$$y = r\sin\theta = \frac{p\sin\theta}{1 + e\cos\theta}. \qquad (1.14)$$

The slope of the tangent to the ellipse at any point is readily calculated to be

$$\frac{dy}{dx} = \frac{\frac{dy}{d\theta}}{\frac{dx}{d\theta}} = -\frac{e + \cos\theta}{\sin\theta}. \qquad (1.15)$$

At $\theta = \pi/2, dy/dx = -e$, which indicates that the angle subtended by the tangent at the end of the latus rectum L is $\tan^{-1}(-e)$, i.e., the angle γ in Fig. 1.2 is $\tan^{-1}e$.

$$\gamma = \sin^{-1}\frac{e}{\sqrt{1 + e^2}} = \cos^{-1}\frac{1}{\sqrt{1 + e^2}} = \tan^{-1}e. \qquad (1.16)$$

This important angle depends on e and is therefore, indicative of the nature of the conic. It is less than, equal to, or greater that $45°$ for the ellipse, the parabola, and the hyperbola, respectively [cf. Van de Kamp (1964)]. This result can also be obtained from analytical geometry [cf. *The Geometry Problem Solver* (1977)].

At the end of the conjugate latus rectum M, one finds, from Sec. 1.3, $dy/dx = e$. The slope of the tangent has a positive sign but the same numerical value as that at L. One can further verify that at the ends of the minor axis (B and C), $dy/dx = 0$, i.e., the tangent is horizontal.

1.6 General Formulation of Particle Dynamics

When regarded as a point particle, a planet's motion consists of both linear motion (*translation*) and angular motion (*revolution* around the Sun). As rigid bodies, planets also possess *rotation* (i.e., *spin*) about their axes. The latter is unrelated to the bodily motion of the planet and is disregarded in this section.

In the general formulation of particle dynamics in three dimensions, the *position vector* $\vec{r}$ from the origin of coordinates marks the instantaneous location of the particle:

$$\vec{r} = r\hat{r}, \tag{1.17}$$

where $\hat{r}$ is the unit vector in the radial direction. The *velocity*, *acceleration* and *jerk* vectors are defined as successive total derivatives of $\vec{r}$ with respect to time:

$$\vec{v} = \frac{d\vec{r}}{dt}, \tag{1.18}$$

$$\vec{a} = \frac{d\vec{v}}{dt} = \frac{d^2\vec{r}}{dt^2}, \tag{1.19}$$

and

$$\vec{j} = \frac{d\vec{a}}{dt} = \frac{d^2\vec{v}}{dt^2} = \frac{d^3\vec{r}}{dt^3}. \tag{1.20}$$

1.7 Kinematics in a Two-Dimensional Plane

For motion in a two-dimensional plane, such as the motion of a planet around the Sun, it is convenient to use plane polar coordinates (r, θ). However, unlike in Cartesian coordinates, the unit vectors in the radial and transverse directions $(\hat{r}, \hat{\theta})$ are not constants, but vary in directions given by [cf. Arfken and Weber (2005)]

$$\frac{d\hat{r}}{d\theta} = \hat{\theta}, \tag{1.21}$$

and

$$\frac{d\hat{\theta}}{d\theta} = -\hat{r}. \tag{1.22}$$

The unit vectors $(\hat{r}, \hat{\theta})$ are related to their Cartesian counterparts $(\hat{x}, \hat{y})$ by the relations [cf. Arfken and Weber (2005)]

$$\hat{r} = \cos\theta\hat{x} + \sin\theta\hat{y}, \tag{1.23}$$

and

$$\hat{\theta} = -\sin\theta\hat{x} + \cos\theta\hat{y}. \tag{1.24}$$

The inverse relations are given by

$$\hat{x} = \cos\theta\hat{r} - \sin\theta\hat{\theta}, \tag{1.25}$$

and

$$\hat{y} = \sin\theta\hat{r} + \cos\theta\hat{\theta}. \tag{1.26}$$

We have, as in three-dimensional space

$$\vec{r} = r\hat{r}. \tag{1.17}$$

By successive differentiations of Eq. (1.17) with time, substitutions from Eqs. (1.21) and (1.22), and simplification, one obtains the general expressions for the velocity and acceleration in plane polar coordinates [cf. Marion and Thornton (1995)]:

$$\vec{v} = v_r\hat{r} + v_\theta\hat{\theta}, \tag{1.27}$$

and

$$\vec{a} = a_r\hat{r} + a_\theta\hat{\theta}, \tag{1.28}$$

where

$$v_r = \frac{dr}{dt},$$ (1.29)

$$v_\theta = r\frac{d\theta}{dt},$$ (1.30)

$$a_r = \frac{d^2r}{dt^2} - r\left(\frac{d\theta}{dt}\right)^2,$$ (1.31)

and

$$a_\theta = r\frac{d^2\theta}{dt^2} + 2\frac{dr}{dt}\frac{d\theta}{dt}.$$ (1.32)

The last term in Eq. (1.31) is recognized as the *centripetal acceleration*, whereas the last term in Eq. (1.32) is the *Coriolis acceleration* [cf. Van de Kamp (1964)].

The *orbital angular velocity* ω and the *orbital angular acceleration* α are given by

$$\omega = \frac{d\theta}{dt},$$ (1.33)

and

$$\alpha = \frac{d\omega}{dt} = \frac{d^2\theta}{dt^2}.$$ (1.34)

The directions of both ω and α are perpendicular to the plane of the orbit, prescribed by the *right-hand rule*.

Finally, the *areal velocity* of the planet in polar coordinates is given by

$$\frac{dA}{dt} = \frac{1}{2}r^2\frac{d\theta}{dt}.$$ (1.35)

Closely-related and proportional to the areal velocity is the *orbital angular momentum* (not to be confused with the *rotational angular momentum*) of the planet

$$l = mr^2 \frac{d\theta}{dt}. \tag{1.36}$$

The constancy of the areal velocity (Kepler's second law) implies the conservation of the angular momentum of the planet. Furthermore, the constancy of either quantity implies that

$$\frac{d}{dt}\left(r^2\frac{d\theta}{dt}\right) = 0. \tag{1.37}$$

This gives, from (1.32): $a_\theta = 0$, which gives us the following theorem.

Theorem 1.1. *If Kepler's law of areas is obeyed in a two-dimensional motion, the acceleration (and therefore force) of the particle has no transverse component. In other words, the conservation of areal velocity (or angular momentum) is a sufficient condition for the force to be **central** in nature, i.e., directed towards the attracting center.*

1.8 Auxiliary Circular Reference Orbit

A circle which circumscribes an ellipse is the *auxiliary circle*. An elliptical orbit and its auxiliary circular orbit have the same major axis. Even though the attracting centers of the two orbits are at different locations, planets describing the two orbits around the same Sun will have the same periods in accordance with the third law of Kepler. Thus, the areal velocities in the two orbits are in the ratio of the areas of the two orbit: $\pi ab : \pi a^2$, i.e., $b : a$.

Figure 1.3 depicts an elliptical planetary orbit with the Sun at the focus F. Also shown in Fig. 1.3 is the auxiliary circular orbit, for which the location of the Sun is at the center O. $\vec{v}_B$ denotes the velocity of the planet at the end of the minor axis B, while $\vec{v}_C$ is the velocity in its auxiliary circular reference orbit. Since $\vec{v}_B$ is parallel to the major axis, the areal velocity of the planet at B is the same

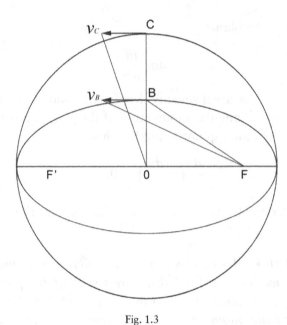

Fig. 1.3

as that referred to O. From the ratios of the two areal velocities, we have

$$\frac{\frac{1}{2}v_B b}{\frac{1}{2}v_C a} = \frac{b}{a},\qquad(1.38)$$

or

$$v_B = v_C.\qquad(1.39)$$

We thus have the following theorem.

Theorem 1.2. *The velocity of a planet at the end of the minor axis is equal to the velocity of the planet in its auxiliary circular orbit [cf. Tan (1979a)]. We shall refer to this velocity as $\bar{v}_0$ in the following section.*

1.9 Speeds at Various Points of an Orbital Ellipse

It is instructive to determine the relative speeds of the planet at specific point on its orbit, e.g., at the ends of the major axis, the minor axis and the latera recta. This can be achieved by geometry using Kepler's law of areas without recourse to formal kinematics and calculus.

In Fig. 1.4, the velocities of the planet at the ends of the major axis are $\vec{v}_P$ and $\vec{v}_A$; that at the end of the minor axis is $\vec{v}_0$; while those at the ends of the latera recta are $\vec{v}_L$ and $\vec{v}_M$. Applying Kepler's second law at the locations P, B and A, we get

$$\frac{1}{2}v_P\overline{PF} = \frac{1}{2}v_0\overline{BO} = \frac{1}{2}v_A\overline{AF}, \tag{1.40}$$

or

$$v_P(1-e) = v_0\sqrt{1-e^2} = v_A(1+e). \tag{1.41}$$

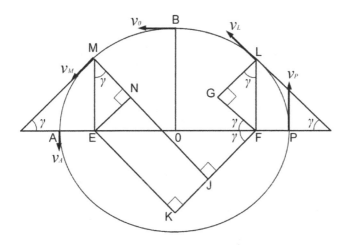

Fig. 1.4

The speeds at the perihelion and aphelion relative to the reference speed v_0 are thus

$$v_P = v_0 \sqrt{\frac{1+e}{1-e}}, \qquad (1.42)$$

and

$$v_A = v_0 \sqrt{\frac{1-e}{1+e}}. \qquad (1.43)$$

The relative speeds at the ends of the major axis follow from Eqs. (1.42) and (1.43):

$$\frac{v_P}{v_A} = \frac{1+e}{1-e}. \qquad (1.44)$$

Furthermore, from Eq. (1.41)

$$v_P v_A = v_0^2. \qquad (1.45)$$

Stated in words, we have the following theorem.

Theorem 1.3. *The speed of a planet at the end of the minor axis is the geometric mean between the speeds at the ends of the major axis* [cf. *Freeman* (1977); *Tan* (1979a)].

In order to find the speeds at the ends of the latera recta, the following constructions are made in Fig. 1.4: Draw $FG \| \vec{v}_L$; and $FK \| \vec{v}_M$; Drop $LG \perp FG$; $MJ \perp FK$; $EK \perp FK$; and $EN \perp MJ$. Then

$$\overline{LG} = \overline{LF} \cos \gamma = a \frac{1 - e^2}{\sqrt{1 + e^2}}, \qquad (1.46)$$

and

$$\overline{MJ} = \overline{MN} + \overline{EK} = \overline{ME} \cos \gamma + \overline{FE} \sin \gamma = a\sqrt{1 + e^2}. \qquad (1.47)$$

Applying Kepler's second law at L, B and M, we get

$$\frac{1}{2} v_L \overline{LG} = \frac{1}{2} v_0 \overline{BO} = \frac{1}{2} v_M \overline{MJ}, \qquad (1.48)$$

or

$$v_L \frac{1 - e^2}{\sqrt{1 + e^2}} = v_0 \sqrt{1 - e^2} = v_M \sqrt{1 + e^2}. \qquad (1.49)$$

The speeds at the ends of the latera recta relative to the reference speed v_0 are thus

$$v_L = v_0 \sqrt{\frac{1 + e^2}{1 - e^2}}, \qquad (1.50)$$

and

$$v_M = v_0 \sqrt{\frac{1 - e^2}{1 + e^2}}. \qquad (1.51)$$

The relative speeds at the ends of the conjugate latera recta follow from Eqs. (1.50) and (1.51):

$$\frac{v_L}{v_M} = \frac{1 + e^2}{1 - e^2}. \qquad (1.52)$$

Furthermore, from Eq. (1.49)

$$v_L v_M = v_0^2. \qquad (1.53)$$

Whence, we have the following theorem.

Theorem 1.4. *The speed of a planet at the end of the minor axis is the geometric mean between the speeds at the ends of the conjugate latera recta* [*cf.* Tan (1979a)].

The speeds of the planet at the special points of the orbit together with the polar coordinates are summarized in Table 1.1.

1.10 General Theorems on Speeds of a Planet

In this section, we shall derive two generalized versions of the velocity theorems discussed in the previous section. The first theorem

Table 1.1. Polar coordinates and speeds of the planet at special points on the orbital ellipse.

Location of Point on Orbit	r	θ	v
Perihelion, end of major axis	$a(1-e)$	0	$v_0\sqrt{\dfrac{1+e}{1-e}}$
End of latus rectum, moving away from Sun	$a(1-e^2)$	$\pi/2$	$v_0\sqrt{\dfrac{1+e^2}{1-e^2}}$
End of minor axis, moving away from Sun	a	$\pi - \cos^{-1}e$	v_0
End of latus rectum through empty focus, moving away from Sun	$a(1+e^2)$	$\pi - \cos^{-1}\dfrac{2e}{1+e^2}$	$v_0\sqrt{\dfrac{1-e^2}{1+e^2}}$
Aphelion, end of major axis	$a(1+e)$	π	$v_0\sqrt{\dfrac{1-e}{1+e}}$
End of latus rectum through empty focus, moving towards Sun	$a(1+e^2)$	$\pi + \cos^{-1}\dfrac{2e}{1+e^2}$	$v_0\sqrt{\dfrac{1-e^2}{1+e^2}}$
End of minor axis, moving towards Sun	a	$\pi + \cos^{-1}e$	v_0
End of latus rectum, moving towards Sun	$a(1-e^2)$	$3\pi/2$	$v_0\sqrt{\dfrac{1+e^2}{1-e^2}}$

gives the product of the speeds of a planet at the ends of any diameter whereas the second theorem gives the ratio of the two speeds. From the two theorems, one can obtain the speed of the planet at any point on its orbit in terms of the reference speed at the end of the minor axis.

In Fig. 1.5, F is the focus occupied by the Sun, E is the empty focus, and OB is the semi minor axis, as usual. IJ is any diameter of the orbital ellipse. $\vec{v}_I$, $\vec{v}_J$ and $\vec{v}_0$ are the velocities of the planet at I, J and B, respectively. Produce the directions of $\vec{v}_J$ and $\vec{v}_I$ to KIS and TJH respectively as shown in the figure. Drop $FG \perp KS$, $FH \perp TH$ and $EK \perp KS$. By virtue of Kepler's second law

$$\frac{1}{2}v_I\overline{FG} = \frac{1}{2}v_J\overline{FH} = \frac{1}{2}v_0\overline{OB}. \tag{1.54}$$

Now there exists an important property of the ellipse (cf. Appendix A.3) which states that the product of the two focal perpendiculars

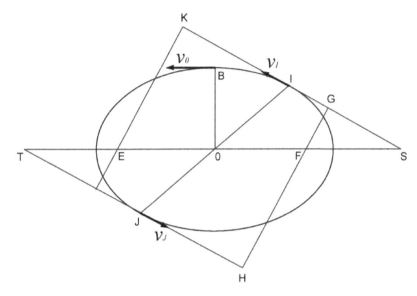

Fig. 1.5

on the tangent at any point I is constant and equal to the square of the semi minor axis [e.g., Salmon (1954) and Lockwood (1961)].

$$\overline{FG} \cdot \overline{EK} = \overline{OB}^2. \tag{1.55}$$

Eliminating $\overline{OB}$ between Eqs. (1.54) and (1.55),

$$v_I v_J = v_0^2. \tag{1.56}$$

Equation (1.56) gives us the following theorem.

Theorem 1.5. *The speed of a planet at the end of the minor axis is equal to the geometric mean between the speeds at the end of any diameter* (*Tan*, 1979b).

Next, from the similar triangles FGS and FHT, we have

$$\frac{\overline{FH}}{\overline{FG}} = \frac{\overline{FT}}{\overline{FS}}. \tag{1.57}$$

Hence from Eq. (1.54):

$$\frac{v_I}{v_J} = \frac{\overline{FH}}{\overline{FG}} = \frac{\overline{FT}}{\overline{FS}}.$$ (1.58)

Thus, we have the next theorem.

Theorem 1.6. *The speeds at the ends of a diameter are inversely proportional to the distances between the focus and the points where the tangents to the ellipse meet the major axis extended* [*Tan*, 1979b].

From Eqs. (1.56) and (1.58), we can find out the speeds at the ends of any diameter in terms of the reference speed v_0 at the end of the minor axis:

$$v_I = v_0 \sqrt{\frac{\overline{FT}}{\overline{FS}}},$$ (1.59)

and

$$v_J = v_0 \sqrt{\frac{\overline{FS}}{\overline{FT}}}.$$ (1.60)

The results of the previous section follow as special cases of the above.

1.11 Velocity Components of a Planet

The formal procedure for obtaining the velocity vector of a planet in its elliptical orbit is to differentiate the position vector with respect to time using the chain rule, the rate of change of the radial unit vector [Eq. (1.21)] and utilize the constancy of the angular momentum vector [Eq. (1.36)]. One can start at the relation

$$\vec{r} = r\hat{r} = \frac{p}{1 + e\cos\theta}\hat{r},$$ (1.61)

whence

$$\vec{v} = \frac{d\vec{r}}{dt} = \frac{dr}{dt}\hat{r} + r\frac{d\hat{r}}{dt} = \frac{dr}{d\theta}\frac{d\theta}{dt}\hat{r} + r\frac{d\hat{r}}{d\theta}\frac{d\theta}{dt}. \tag{1.62}$$

Upon carrying out the differentiation, we get

$$\vec{v} = Ve\sin\theta\hat{r} + V(1 + e\cos\theta)\hat{\theta}, \tag{1.63}$$

where

$$V = \frac{l}{mp}. \tag{1.64}$$

The velocities at the perihelion and aphelion points readily follow from Eq. (1.63) by setting $\theta = 0$ and $\theta = \pi$, respectively:

$$\vec{v}_P = (1 + e)V\hat{\theta}, \tag{1.65}$$

and

$$\vec{v}_A = (1 - e)V\hat{\theta}. \tag{1.66}$$

Both of these velocities have no radial components. The magnitudes of the two velocities (i.e., the speeds) are simply

$$v_P = (1 + e)V, \tag{1.67}$$

and

$$v_A = (1 - e)V. \tag{1.68}$$

The velocity at the end of the latus rectum (L of Fig. 1.2) is likewise obtained from Eq. (1.63) by putting $\theta = \pi/2$:

$$\vec{v}_L = Ve\hat{r} + V\hat{\theta}, \tag{1.69}$$

which gives the speed at L

$$v_L = \sqrt{1 + e^2}\,V. \tag{1.70}$$

The velocity at the end of the minor axis (B of Fig. 1.2) is obtained by putting $\theta = \pi - \alpha$ in Eq. (1.63), where α is given by (1.11):

$$\vec{v}_B = Ve\sqrt{1 - e^2}\,\hat{r} + V(1 - e^2)\hat{\theta}. \tag{1.71}$$

The speed of the planet at the end of the minor axis is then

$$v_B = \sqrt{1 - e^2}\, V. \qquad (1.72)$$

The velocity at one other special point is that at M (Fig. 1.2). We get by putting $\theta = \pi - \beta$ in Eq. (1.63) with β given by Eq. (1.12):

$$\vec{v}_M = \frac{1 - e^2}{1 + e^2} V e \hat{r} + \frac{1 - e^2}{1 + e^2} V \hat{\theta}. \qquad (1.73)$$

The speed of the planet at M is then

$$v_M = \frac{1 - e^2}{\sqrt{1 + e^2}} V. \qquad (1.74)$$

The velocity components at L', B' and M' can likewise be calculated. The results are summarized in Table 1.2. They are also displayed in Fig. 1.6. One can further verify that the velocity

Table 1.2. Velocity components of the planet at special points on the orbital ellipse.

Location of Point on Ellipse	v_r	v_θ	v
Perihelion, end of major axis	0	$V(1 + e)$	$V(1 + e)$
End of latus rectum, moving away from Sun	Ve	V	$V\sqrt{1 + e^2}$
End of minor axis, moving away from Sun	$Ve\sqrt{1 - e^2}$	$V(1 - e^2)$	$V\sqrt{1 - e^2}$
End of latus rectum through empty focus, moving away from Sun	$Ve\dfrac{1 - e^2}{1 + e^2}$	$V\dfrac{1 - e^2}{1 + e^2}$	$V\dfrac{1 - e^2}{\sqrt{1 + e^2}}$
Aphelion, end of major axis	0	$V(1 - e)$	$V(1 - e)$
End of latus rectum through empty focus, moving towards Sun	$-Ve\dfrac{1 - e^2}{1 + e^2}$	$V\dfrac{1 - e^2}{1 + e^2}$	$V\dfrac{1 - e^2}{\sqrt{1 + e^2}}$
End of minor axis, moving towards Sun	$-Ve\sqrt{1 - e^2}$	$V(1 - e^2)$	$V\sqrt{1 - e^2}$
End of latus rectum, moving towards Sun	$-Ve$	V	$V\sqrt{1 + e^2}$

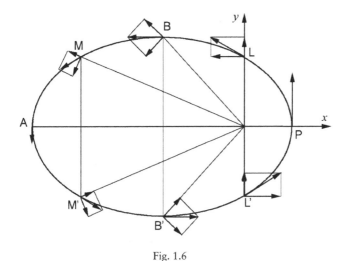

Fig. 1.6

theorems obtained earlier hold:

$$v_P v_A = v_B^2,$$ (1.75)

and

$$v_L v_M = v_B^2.$$ (1.76)

1.12 Planetary Motion as the Sum of Circular and Linear Motions

Quite interestingly, Eq. (1.63) can be rewritten in the following form

$$\vec{v} = Ve(\sin\theta\hat{r} + \cos\theta\hat{\theta}) + V\hat{\theta}.$$ (1.77)

The first term on the right-hand side of Eq. (1.77) is recognized as a constant velocity vector of magnitude Ve directed along the positive y-direction ($\theta = \pi/2$), whereas the second term is a uniformly rotating vector in the counter-clockwise direction of magnitude V [cf. Van de Kamp (1967)]. Thus, we can state the following theorem.

Theorem 1.7. *The motion of a planet in an elliptical orbit is a superposition of a uniform circular motion and a constant linear motion along the positive minor axis. In the case of $e = 0$, the constant velocity term is zero, and we are left with a uniform circular motion.*

By virtue of Eq. (1.26), we can recast Eq. (1.77) in a simple form in terms of mixed Cartesian and polar coordinates:

$$\vec{v} = Ve\hat{y} + V\hat{\theta}. \tag{1.78}$$

1.13 Orbit of the Earth and its Eccentricity

The orbit of the Earth is slightly elliptical having an eccentricity of 0.017 [cf. McBride and Gilmour (2003)]. The time of one revolution of the Earth around the Sun (i.e., its period) is 365.25 days [cf. McBride and Gilmour (2003)]. The orbital plane of the Earth is called the *ecliptic*. The Earth's perihelion falls most frequently around January 3 and its aphelion falls most frequently around July 5. Unrelated to the revolution is the rotation of the Earth about its own axis or its spin. The tilt of the spin axis from the normal to the ecliptic of 23.5° is responsible for the seasons. The *winter solstice* marks the day when the northern hemisphere is tilted farthest away from the Sun and occurs most frequently on December 21. The *summer solstice* falls most frequently on June 21. On that day, the southern hemisphere is tilted farthest away from the Sun. Two other days of importance are the *vernal equinox* and the *autumnal equinox* on which days and nights are equal everywhere on the globe. The former occurs most frequently on March 20 while the latter occurs most frequently on September 22 and 23.

In Fig. 1.7, P, A, W, S, E and E' mark the perihelion, aphelion, winter solstice, summer solstice, vernal equinox and autumnal equinox, respectively. Also shown in the figure is the latus rectum at the focus LOL'. It is evident that the angles WOP, EOL, SOA

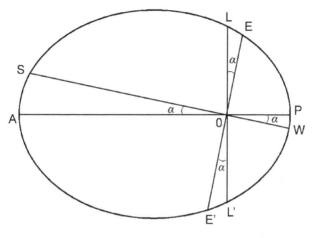

Fig. 1.7

and $E'OL'$ are all equal (to α, say). It is also apparent that α is coincidentally quite small. The interval from W to P is only 13 days, while the interval from S to A is just 14 days. Since the Earth sweeps 360° around the Sun in 365.25 days, these two intervals subtend angles of approximately 12.8° and 13.8°, respectively, at the center of the ellipse, or approximately 13.3° at the focus.

The difference between the intervals of passage from W to P and from S to A is due to the fact that the planet is fastest at P and slowest at A. Interestingly, this provides us with a means to determine the eccentricity of the Earth (Tan, 2004–2005). In accordance with Kepler's second law, we can write

$$\frac{1}{2}r_P^2\frac{d\theta_P}{dt_P} = \frac{1}{2}r_A^2\frac{d\theta_A}{dt_A}. \tag{1.79}$$

Setting $r_P = a(1 - e); r_A = a(1 + e); d\theta_P = d\theta_A = \alpha; dt_P = 13$ days; and $dt_A = 14$ days, we obtain

$$\frac{a^2(1 - e^2)}{a^2(1 + e^2)} = \frac{13}{14}, \tag{1.80}$$

which gives

$$e = \frac{1 - \sqrt{\frac{13}{14}}}{1 + \sqrt{\frac{13}{14}}} \approx 0.0185. \qquad (1.81)$$

This agrees roughly with the actual value of the Earth's eccentricity of 0.017.

1.14 Bode's Law

In our solar system, a common unit of distance is the **Astronomical Unit** (AU), which is defined as the average distance of the Earth from the Sun. The locations of the planets follow a curious pattern prescribed by what is commonly known as **"Bode's Law"** (*sic*). It was actually discovered by Titius in 1766 [cf. Murray and Dermott (1999)], but propagated by Bode in 1772 [cf. Bate *et al.* (1971)]. According to this "law", the distances of the planets from the Sun in AU are approximately given by the following empirical relation [cf. Murray and Dermott (1999)]

$$d = \frac{3 \times 2^n + 4}{10}, \qquad (1.82)$$

where $n = -\infty, 0, 1, 2, 4$ and 5 correspond to the known planets at that time, viz., Mercury, Venus, Earth, Mars, Jupiter and Saturn, respectively. Bode's Law correctly predicted the locations of a missing planet (corresponding to $n = 3$) at $d = 2.8$ AU and a post-Saturnian planet ($n = 6$) at $d = 19.6$ AU. The subsequent discoveries of Uranus in 1781 and (asteroid) Ceres in 1801 were considered the greatest triumphs of this empirical law.

The existence of a similar law (in orbital periods) for the satellites of Uranus was uncovered by Dermott (1973).

Exercises

1.1. Find the period of Mars in Earth days, given that it orbits the Sun at 1.52 AU.

1.2. Derive Eqs. (1.3) and (1.4).

1.3. Verify Eqs. (1.6) to (1.10).

1.4. Derive Eqs. (1.11) and (1.12).

1.5. Derive Eqs. (1.21) and (1.22).

1.6. Verify Eqs. (1.23) to (1.26).

1.7. Derive Eqs. (1.31) and (1.32).

1.8. Give a geometrical derivation for the expression of areal velocity (1.35).

1.9. Verify Eqs. (1.45) and (1.63).

1.10. Verify the entries of Table 1.2.

1.11. Show that $\hat{y} = \sin\theta\hat{r} + \cos\theta\hat{\theta}$ is a unit vector along the positive y-axis.

1.12. Calculate the velocity components and speeds of the planet at the ends of the minor axis and the latera recta (B', L' and M' of Fig. 1.6) during the descending phase of the planet's motion.

1.13. Compare the distances of the planets given by Bode's Law with their actual distances.

Newton's Law of Gravitation

2.1 Gravity and Gravitation

Gravity refers to the force of attraction of an object by the Earth, whereas *gravitation* is the force of attraction between any two objects in the universe. Gravity dictates the motion of falling objects and projectiles on the Earth's surface. It also maintains the motion of the moon and artificial satellite around the Earth. Gravity is simply a special case of gravitational force attributed to the Earth.

Gravitation is by far the *weakest force* in nature. However, unlike the short-range nuclear forces, it has a far longer range. Furthermore, unlike the electromagnetic force, it does not cancel out but adds up to a substantial force for large objects. Thus, the motion of the heavenly bodies in the universe is governed by gravitation. Gravitational force is always *attractive* in nature and no repulsive gravity has been observed. The attraction between two objects is mutual and *Newton's third law of motion* is obeyed.

2.2 Newton's Law of Gravitation

Sir Isaac Newton deduced his *Law of Universal Gravitation* from Kepler's laws of planetary motion. This law, along with his three laws of motion, was finally published, at the behest of his friends, in his *Principia* in 1687, some two decades after its discovery. The following is an uncorrupted version of this law: every body (a mass) attracts every other body (another mass) in the universe with a force which is directly proportional to each mass and indirectly proportional to the square of the distance between the two. The magnitude of the gravitational force between two masses m_1 and m_2, separated by a distance r is given by

$$f = G\frac{m_1 m_2}{r^2}, \qquad (2.1)$$

where G, the *universal gravitational constant* is 6.67×10^{-8} c.g.s. units (or 6.67×10^{-11} SI units).

Newton's law of gravitation is a fundamental law of nature which governs the motion of heavenly bodies. All three Kepler's laws of planetary motion, and much more, follow from Newton's law, even though the formal derivations require *calculus*, which Newton himself invented as a necessity. The Keplerian problem is applicable when the two following conditions are met: (i) The distance between the Sun and the planet is so much greater than the dimensions of the Sun and the planet such that they can be approximated as point particles; and (ii) The mass of the planet is negligible compared with that of the Sun, so that the latter remains basically stationary. Both of these conditions are substantially met for planetary motion in our solar system.

2.3 Centripetal Force and Radius of Curvature

In a uniform circular motion, the *centripetal force* on an object is always directed towards the center. The magnitude of this force on a mass m moving in a circle of radius r with a velocity $\bar{v}$ is given by the well-known expression

$$f_c = m\frac{v^2}{r}. \tag{2.2}$$

This result, which can be deduced from vectorial kinematics, is originally credited to Huygens (1673). For a general curvilinear motion, r at any point on the curved path is replaced by the *radius of curvature* ρ at that point, which is the radius of the circle (called the *circle of curvature*) of which the segment of the path at that point is a part

$$f_c = m\frac{v^2}{\rho}. \tag{2.3}$$

Since a circle can be drawn passing through any three non-colinear points, the circle of curvature is defined by the point on the curve under consideration plus two other neighboring points on either side of it.

The radius of curvature in polar coordinates is readily found in the literature [cf. Gellert *et al.* (1977)]:

$$\rho = \frac{\left[r^2 + \left(\frac{dr}{d\theta}\right)^2\right]^{3/2}}{r^2 + 2\left(\frac{dr}{d\theta}\right)^2 - r\frac{d^2r}{d\theta^2}}. \tag{2.4}$$

The radius of curvature at any point of the Keplerian ellipse can be evaluated from Eq. (1.5). We have

$$r = \frac{p}{1 + e\cos\theta}, \tag{2.5}$$

$$\frac{dr}{d\theta} = \frac{pe\sin\theta}{(1 + e\cos\theta)^2}, \tag{2.6}$$

and

$$\frac{d^2r}{d\theta^2} = \frac{pe^2 + pe^2\sin^2\theta + pe\cos\theta}{(1 + e\cos\theta)^3}. \tag{2.7}$$

Substituting Eqs. (2.5)–(2.7) in (2.4) and simplifying, we arrive at

$$\rho = \frac{p(1 + e^2 + 2e\cos\theta)^{3/2}}{(1 + e\cos\theta)^3}. \tag{2.8}$$

The radii of curvature at familiar points on the orbital ellipse readily follow from Eq. (2.8). At the ends of the major axis, i.e., at the perihelion P ($\theta = 0$) and aphelion $A(\theta = \pi)$, we have

$$\rho_P = p = \rho_A, \tag{2.9}$$

or, the radii of curvature at the ends of the major axis are just equal to the semi latus rectum p. At the ends of the latus rectum $L(\theta = \pi/2)$ and $L'(\theta = 3\pi/2)$,

$$\rho_L = p(1 + e^2)^{3/2} = \rho_{L'}. \tag{2.10}$$

At the ends of the minor axis $B(\theta = \pi - \alpha)$ and $C(\theta = \pi + \alpha)$ with $\alpha = \cos^{-1}(-e)$ [Eq. (1.11)], we have

$$\rho_B = \frac{p}{(1 - e^2)^{3/2}} = \rho_C. \tag{2.11}$$

Likewise, at the ends of the latus rectum through the empty focus $M(\theta = \pi - \beta)$ and $M'(\theta = \pi + \beta)$ with β given by Eq. (1.12), we get

$$\rho_M = p(1 + e^2)^{3/2} = \rho_{M'}. \tag{2.12}$$

In terms of semi major axes a and b, one can verify the following familiar results

$$\rho_P = \rho_A = \frac{b^2}{a}, \tag{2.13}$$

and

$$\rho_B = \frac{a^2}{b}. \tag{2.14}$$

2.4 Kepler's Laws and Newton's Law for Circular Orbits

According to Newton's law, the gravitational force diminishes with the inverse square of the distance, much like the intensity of light from a point source or the loudness of sound, and thus appears quite natural in that sense. A century later, Coulomb would verify a similar law for the electrical interaction between two charges. It is an easy matter to derive the inverse square law from Kepler's laws for circular orbits and vice-versa.

First, one can easily derive the inverse square law from circular planetary orbits. The circle is the limiting case of an ellipse, and circular orbits are legitimate Keplerian orbits in accordance with Kepler's first law. Kepler's second law ensures that the speed of a planet in a circular orbit of radius r is constant and equal to

$$v = \frac{2\pi r}{P}, \tag{2.15}$$

where the period P is given by Kepler's third law. With a constant of proportionality k, we have from Eq. (1.1)

$$P^2 = kr^3. \tag{2.16}$$

Upon substitution of Eqs. (2.15) and (2.16) in the expression for the centripetal force (2.2), we obtain the desired result:

$$f = \frac{4\pi^2}{k}\frac{1}{r^2}. \tag{2.17}$$

Next, one can also arrive at Kepler's third law from Newton's law of gravitation as applied to circular orbits. Here

$$P = \frac{2\pi r}{v}. \tag{2.18}$$

Letting M be the mass of the sun and m that of the planet in Eq. (2.1),

$$f = G\frac{Mm}{r^2}. \tag{2.19}$$

Equating (2.2) with (2.19) gives the speed of the planet in the circular orbit

$$v = \sqrt{\frac{GM}{r}}, \qquad (2.20)$$

whence from (2.18)

$$P = \frac{2\pi}{\sqrt{GM}} r^{3/2}, \qquad (2.21)$$

which is Kepler's third law.

2.5 Newton's Law of Gravitation from Elliptical Orbits

The validity of Newton's inverse square law of gravitation has been demonstrated at special points of an elliptical orbit, namely the apsidal points of perihelion and aphelion (Macklin, 1971) and the end of the semi-minor axis (Weinstock, 1972). In Fig. 2.1, F is the focus, P the perihelion and A the aphelion of the planet's orbit. At P and A, the centripetal forces are directed towards the focus and are therefore equal to the gravitational forces. From Eq. (2.3), we

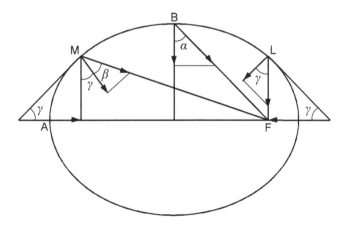

Fig. 2.1

have

$$f_P = \frac{mv_P^2}{\rho_P}, \tag{2.22}$$

and

$$f_A = \frac{mv_A^2}{\rho_A}. \tag{2.23}$$

To verify Newton's law, it is sufficient to show that

$$f_P r_P^2 = f_A r_A^2, \tag{2.24}$$

where r_P and r_A are the apsidal distances from the Sun. With substitutions from Eqs. (2.22), (1.6), (1.42) and (2.9), we get

$$f_P r_P^2 = mv_0^2 a. \tag{2.25}$$

Likewise from Eqs. (2.23), (1.7), (1.43) and (2.9)

$$f_A r_A^2 = mv_0^2 a. \tag{2.26}$$

Thus, relation (2.24) is satisfied.

It is instructive to investigate the validity of Newton's law of gravitation at other points on the orbit where the centripetal force is not radially inwards, for example, at the ends of the minor axis (B) and latera recta (L and M). One needs to remember that the centripetal force is nothing but the component of the gravitational force along the normal to the trajectory. At B, we have (vide Fig. 2.1):

$$f_B = \frac{mv_B^2}{\rho_B} \sec \alpha, \tag{2.27}$$

where

$$\sec \alpha = \frac{a}{b}. \tag{2.28}$$

Thus

$$f_B r_B^2 = \frac{mv_B^2 r_B^2}{\rho_B} \frac{a}{b}. \tag{2.29}$$

Inserting the values for v_B and ρ_B from Eqs. (1.39) and (2.14), we get (Weinstock, 1972)

$$f_B r_B^2 = m v_0^2 a, \tag{2.30}$$

which is in agreement with Eqs. (2.25) and (2.26).

At L, we have, similarly (see Fig. 2.1)

$$f_L = \frac{m v_L^2}{\rho_L} \sec \gamma, \tag{2.31}$$

where, from Eq. (1.16)

$$\sec \gamma = \sqrt{1 + e^2}. \tag{2.32}$$

Thus

$$f_L r_L^2 = \frac{m v_L^2 r_L^2}{\rho_L} \sqrt{1 + e^2}. \tag{2.33}$$

Substituting the values of v_L and ρ_L from Eqs. (1.50) and (2.10), we arrive at

$$f_L r_L^2 = m v_0^2 a. \tag{2.34}$$

Finally, at M, we have (vide Fig. 2.1)

$$f_M = \frac{m v_M^2}{\rho_M} \sec \beta. \tag{2.35}$$

By the *reflective property* of the ellipse (cf. Appendix A.3), $\beta = \gamma$, whence by Eq. (2.32)

$$\sec \beta = \sec \gamma = \sqrt{1 + e^2}. \tag{2.36}$$

Thus

$$f_M r_M^2 = \frac{m v_M^2 r_M^2}{\rho_M} \sqrt{1 + e^2}. \tag{2.37}$$

With substitutions for v_M and ρ_M from Eqs. (1.51) and (2.12), we once again obtain

$$f_M r_M^2 = m v_0^2 a. \tag{2.38}$$

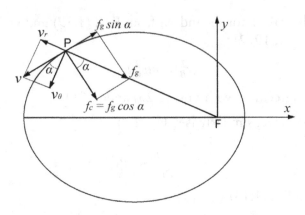

Fig. 2.2

The inverse square law can further be established for any general point P on the elliptic orbit (Fig. 2.2). We begin with the polar equation of the ellipse [Eq. (2.5)] and utilize the fact that the angular momentum of the planet l is a constant:

$$l = mr^2 \frac{d\theta}{dt}.$$
(2.39)

The radial and transverse components of the velocity vector readily follow

$$v_r = \frac{dr}{dt} = \frac{dr}{d\theta}\frac{d\theta}{dt} = \frac{l}{mp}e\sin\theta,$$
(2.40)

and

$$v_\theta = r\frac{d\theta}{dt} = \frac{l}{mp}(1 + e\cos\theta).$$
(2.41)

Squaring and adding gives [cf. Tan (2004)]

$$v^2 = v_r^2 + v_\theta^2 = \frac{l^2}{m^2p^2}(1 + e^2 + 2e\cos\theta),$$
(2.42)

and

$$v = \frac{l}{mp}\sqrt{1 + e^2 + 2e\cos\theta}.$$
(2.43)

Substituting from the radius of curvature [Eq. (2.8)], we get for the centripetal force at P,

$$f_c = \frac{mv^2}{\rho} = \frac{l^2}{mp^3} \frac{(1 + e\cos\theta)^3}{\sqrt{1 + e^2 + 2e\cos\theta}}. \qquad (2.44)$$

If f_g is the gravitational force, we have from (2.27) (vide Fig. 2.2):

$$f_g = f_c \sec\alpha = f_c \frac{v}{v_\theta}. \qquad (2.45)$$

Substituting Eqs. (2.41), (2.43) and (2.44) in (2.45), we finally arrive at the inverse square law:

$$f_g = \frac{l^2}{mp} \frac{1}{r^2} \propto \frac{1}{r^2}. \qquad (2.46)$$

2.6 The Nature of Gravitational Force

In three-dimensional space, Newton's law of gravitation as applied to the Sun M and its planet m can be written in *spherical polar coordinates* with the Sun at the origin as thus

$$\vec{f} = -G\frac{Mm}{r^2}\hat{r} = -G\frac{Mm}{r^3}\vec{r}, \qquad (2.47)$$

where

$$\vec{r} = r\hat{r}. \qquad (2.48)$$

is the position vector of the planet and $\hat{r}$ is the unit vector along the radial direction. The negative sign indicates that gravitation is an *attractive force* by nature. Clearly the magnitude of this force is independent of the direction and is therefore *isotropic*. Being independent of the angular coordinates, it is also *spherically symmetric*. Furthermore, being always directed towards a point, it is a *central force*. Other examples of the central force include the dipole, quadrupole and multipole forces, the elastic force and Van der Waals' force.

Like all central forces, the gravitational force is an *irrotational vector*. The latter is defined as any vector whose curl vanishes:

$$\vec{\nabla} \times \vec{f} = \vec{0}, \tag{2.49}$$

where $\vec{0}$ is the null vector. It is easy to show from the second form of $\vec{f}$ in Eq. (2.47) that it satisfies the relation (2.49). Unlike most other central forces, however, the inverse square central force is also a *solenoidal vector*, which is defined as any vector whose divergence vanishes:

$$\vec{\nabla} \cdot \vec{f} = 0. \tag{2.50}$$

Once again, the second form of $\vec{f}$ in Eq. (2.47) facilitates the validation of Eq. (2.50).

Like all translational vectors (e.g., the position, velocity, momentum and acceleration vectors) and unlike rotational vectors (e.g., angular displacement, angular velocity, angular momentum, angular acceleration and torque vectors), the gravitational force changes sign under the *inversion of coordinates*.

$$\vec{f}(-\vec{r}) = -\vec{f}(\vec{r}). \tag{2.51}$$

It is therefore a *true vector* or *polar vector* [cf. Arfken and Weber (2005)].

Unlike friction and air resistance (which are dissipative forces), gravitational force is *conservative*. A conservative force is obtained as the negative gradient of a scalar *potential energy* V:

$$\vec{f} = -\vec{\nabla} V. \tag{2.52}$$

The *gravitational potential energy* of a mass at any point in a gravitational field is defined as the work done in bringing it from infinity up to that point:

$$V = \int_{\infty}^{r} \vec{f} \cdot d\vec{r}. \tag{2.53}$$

Alternatively and equivalently, it has also been defined as the work done in removing the mass from that point to infinity. One can arrive at the gravitational potential energy of m at r using either definition. Putting Eq. (2.47) in (2.53), we obtain:

$$V(r) = -G\frac{Mm}{r}. \tag{2.54}$$

Evidently, for a conservative force, relation (2.49) is satisfied:

$$\vec{\nabla} \times \vec{f} = -\vec{\nabla} \times \vec{\nabla}V = \vec{0}. \tag{2.55}$$

Finally, the net work done in moving m along a closed path is zero:

$$\oint \vec{f} \cdot d\vec{r} = -\oint \vec{\nabla}V \cdot d\vec{r} = -\oint dV = 0, \tag{2.56}$$

since V is a single-valued function of position. Thus we have three equivalent definitions of a conservative force given by Eqs. (2.52), (2.55) and (2.56) [cf. Arfken and Weber (2005)].

2.7 General Dynamics of a Particle

Consider the general motion of a particle of mass m moving with a velocity $\vec{v}$. The **linear momentum** of the particle is the product of m and $\vec{v}$, which is a vector in the direction of the velocity:

$$\vec{p} = m\vec{v} = m\frac{d\vec{r}}{dt}. \tag{2.57}$$

The momentum represents the quantity of motion and was quite appropriately called "motion" once upon a time. The **translational**

kinetic energy of the particle (a scalar) is defined as:

$$T = \frac{1}{2}m\vec{v} \cdot \vec{v} = \frac{1}{2}mv^2. \tag{2.58}$$

Newton's first law of motion states that for force-free motion, the velocity and therefore momentum of the particle remain constant. **Newton's second law of motion** states that the rate of change of momentum of the particle is equal to the **external applied force**:

$$\vec{f} = \frac{d\vec{p}}{dt} = m\frac{d\vec{v}}{dt} = m\vec{a}. \tag{2.59}$$

The **elementary work done** by the force in causing a displacement $d\vec{r}$ is given by the scalar product

$$dW = \vec{f} \cdot d\vec{r}. \tag{2.60}$$

Substituting from Eqs. (1.18) and (1.19), one obtains

$$dW = m\frac{d\vec{v}}{dt} \cdot \vec{v}dt = m\vec{v} \cdot d\vec{v} = \frac{1}{2}md(\vec{v} \cdot \vec{v}) = d\left(\frac{1}{2}m\vec{v} \cdot \vec{v}\right) = dT. \tag{2.61}$$

The total work done in moving the mass from positions 1 to 2 is given by the integral

$$W = \int_1^2 dW = [T]_1^2 = T_2 - T_1. \tag{2.62}$$

This is the **work-energy theorem** which states that the work done by the external force goes to increasing the kinetic energy of the particle [cf. Marion and Thornton (1995)]. One should note that the convention of work done by an enclosed gas in thermodynamics is the opposite to that of particle dynamics.

Since work done results in the decrease of potential energy, we can write

$$dW = -dV. \qquad (2.63)$$

Comparing with Eq. (2.61), we have

$$dT + dV = 0, \qquad (2.64)$$

which when integrated gives

$$T + V = E = const. \qquad (2.65)$$

This is the *law of conservation of energy* as applied to the orbiting planet [cf. Van de Kamp (1964)].

An example in planetary motion is relevant. When the planet descends towards the sun, the work done by the gravitational force goes to increasing the translational kinetic energy, and hence its velocity. In the ascending mode, however, the force and displacement are oppositely directed: the work done is negative, and the planet loses its kinetic energy and velocity. The total energy of the planet, however, remains constant due to the conservative nature of the force.

2.8 Kepler's Third Law as an Extension of the First Two

It has been demonstrated that Kepler's third law follows from the first two if either energy or centripetal force considerations are made (Weinstock, 1962). The first demonstration assumes the expressions for the kinetic and potential energies of the planet plus the law of conservation of energy. Refer to the perihelion and aphelion points of the orbit in Fig. 2.1. Kepler's second law as applied to P and A

gives

$$\frac{1}{2}a(1-e)v_P = \frac{1}{2}a(1+e)v_A = \frac{\pi ab}{P}. \tag{2.66}$$

From Eq. (2.66), it follows that

$$v_P = \frac{2\pi a}{P}\sqrt{\frac{1+e}{1-e}}, \tag{2.67}$$

and

$$v_A = \frac{2\pi a}{P}\sqrt{\frac{1-e}{1+e}}. \tag{2.68}$$

Applying the law of conservation of energy at P and A, we have

$$\frac{1}{2}mv_P^2 - \frac{GMm}{a(1-e)} = \frac{1}{2}mv_A^2 - \frac{GMm}{a(1+e)}. \tag{2.69}$$

Substituting for v_P and v_A from Eqs. (2.67) and (2.68) and solving for the period P, one obtains Kepler's third law:

$$P^2 = \frac{4\pi^2}{GM}a^3. \tag{2.70}$$

The second demonstration equates the expressions for the gravitational force of Newton and the centripetal force of Huygens. We have at P and A the following:

$$\frac{GMm}{[a(1-e)]^2} = \frac{mv_P^2}{\rho}, \tag{2.71}$$

and

$$\frac{GMm}{[a(1+e)]^2} = \frac{mv_A^2}{\rho}. \tag{2.72}$$

With substitutions for v_P and v_A from Eqs. (2.67) and (2.68) and the expression for ρ from Eq. (2.9), both of the Eqs. (2.71) and (2.72) give the desired result (2.70).

2.9 The Equation of Energy

We had earlier derived the expression for the square of the velocity in Keplerian motion:

$$v^2 = \frac{l^2}{m^2 p^2}(1 + e^2 + 2e\cos\theta). \tag{2.33}$$

With substitutions from Eqs. (1.4) and (1.5), the above equation can be recast into the form

$$v^2 = \frac{l^2}{m^2 p}\left(\frac{2}{r} - \frac{1}{a}\right). \tag{2.73}$$

This is a form of the **Equation of Energy** [cf. Van de Kamp (1964) and Danby (1988)], formerly known as the **Vis Viva Integral** [cf. Moulton (1970) and Brouwer and Clemence (1961)]. To obtain the standard form of the energy equation, we compare the force equations at special points of the orbit. At the perihelion P and the aphelion A, the centripetal force is radial and equal to the gravitational force:

$$\frac{mv^2}{\rho} = \frac{GMm}{r^2}. \tag{2.74}$$

Putting the values of ρ and r at either P or A, we get from Eqs. (2.73) and (2.74):

$$\frac{l^2}{m^2 p} = GM. \tag{2.75}$$

Thus, the standard form of the equation of energy is

$$v^2 = GM\left(\frac{2}{r} - \frac{1}{a}\right). \tag{2.76}$$

Equation (2.75) is widely used in orbital mechanical calculations. The product GM is called the **gravitational parameter** for the attracting mass and is commonly denoted by μ.

Two special is values of the velocity are obtained from Eq. (2.76). Firstly, for a circular orbit, $r = a$, and we have the *circular velocity* v_c,

$$v_c = \sqrt{\frac{GM}{a}} = \sqrt{\frac{GM}{r}}. \qquad (2.77)$$

Secondly, for parabolic motion, $a \to \infty$. Then we have the *parabolic velocity* v_p or *velocity of escape* v_e,

$$v_p = v_e = \sqrt{\frac{2GM}{r}} = \sqrt{2}v_c. \qquad (2.78)$$

The parabolic velocity is square root of two times the circular velocity [cf. Van de Kamp (1964)].

An interesting observation follows from the equation of energy [Eq. (2.76)]. For any orbit with a given a, the orbital velocity of the planet depends solely on the radial distance r of the planet from the Sun. The orbital speed v at any point of the elliptical orbit is actually the speed acquired by an object in a free fall from a circular orbit of radius $2a$ centered at the Sun [Fig. 2.3, cf. Van de Kamp (1964)].

The following furnishes the proof. From Newton's second law of motion and his law of gravitation, we can write the force equation

$$m\frac{dv}{dt} = -\frac{GMm}{r^2}. \qquad (2.79)$$

It is a common technique in mechanics to convert the time variable to the space variable. The left-hand side of Eq. (2.79) is converted to

$$m\frac{dv}{dr}\frac{dr}{dt} = mv\frac{dv}{dr}. \qquad (2.80)$$

Equation (2.79) can now be integrated from the radial distance of $2a$ up to r:

$$m\int_0^v vdv = -\int_{2a}^r \frac{GMm}{r^2}dr. \qquad (2.81)$$

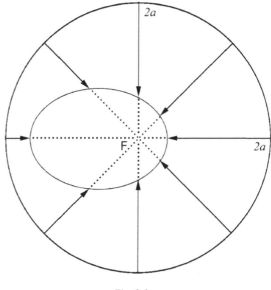

Fig. 2.3

Upon carrying out the integrations, we arrive at Eq. (2.76). Hence proved.   □

2.10 Energy of the Orbit

By putting the value of v^2 from the energy equation (2.76) into the law of conservation of energy (2.65), one arrives at the *energy of the orbit*

$$T + V = E = -\frac{GMm}{2a}. \tag{2.82}$$

Two observations are made from Eq. (2.82). Firstly, the total energy of the planet is negative. This signifies that the planet is in a *bound state* around the Sun much like an electron in an atom. In order to free the planet from the gravitational field of the Sun, an amount of energy equal to E in Eq. (2.82) must be added to the planet.

Secondly, the energy of orbit is the same for all orbits having the same semi-major axis *a*, i.e., the energy is independent of the minor axis, or the eccentricity of the orbit, for that matter. Coupled with Kepler's harmonic law, we have the following theorem.

Theorem 2.1. *Planetary orbits with the same semi-major axes have the same period and energy.*

If ρ is the perpendicular distance from the Sun to the extension of the velocity vector (ρ is not to be confused with the radius of curvature, which was expressed by the same symbol), then it serves as the lever arm of the angular momentum of the planet. Thus

$$mv\rho = l. \tag{2.83}$$

Substitution of Eq. (2.83) in the energy equation (2.82) renders it to the form [cf. Lockwood (1967)]

$$\frac{a}{GMm^2}\frac{l^2}{\rho^2} = \frac{2a}{r} - 1, \tag{2.84}$$

which is the pedal form of the equation of the orbital ellipse (cf. Appendix A.11).

The special case of the circular orbit calls for a little attention. For $r = a$, we have

$$T = \frac{GMm}{2a}, \tag{2.85}$$

$$V = -\frac{GMm}{a}, \tag{2.86}$$

and

$$E = -\frac{GMm}{2a}. \tag{2.87}$$

Thus,

$$T = -\frac{V}{2} = -E. \tag{2.88}$$

Equation (2.88) is a special case of the "virial theorem" discussed in the following section.

2.11 The Virial Theorem of Clausius

Consider the function given by the scalar product of the momentum and position vectors:

$$G = \vec{p}.\vec{r}. \tag{2.89}$$

Its time derivative is given by

$$\frac{dG}{dt} = \vec{p}.\frac{d\vec{r}}{dt} + \frac{d\vec{p}}{dt}.\vec{r} = m\vec{v}.\vec{v} + \vec{f}.\vec{r}. \tag{2.90}$$

When the force is given by Newton's law of gravitation (2.47), Eq. (2.90) reduces to

$$\frac{dG}{dt} = 2T + V. \tag{2.91}$$

The time average of the left-hand side is given by

$$\frac{\int\limits_0^\tau \frac{dG}{dt}dt}{\int\limits_0^\tau dt} = \frac{G(\tau) - G(0)}{\tau}. \tag{2.92}$$

Since for a finite system, the value of G is always finite, the right-hand side of Eq. (2.92) becomes zero after a sufficiently long interval τ. Hence, we get from Eq. (2.91):

$$2\langle T \rangle + \langle V \rangle = 0, \tag{2.93}$$

or

$$\langle T \rangle = -\frac{\langle V \rangle}{2}. \tag{2.94}$$

This is the general case of the *virial theorem* (2.88) for a planetary orbit of any shape or size.

Exercises

2.1. Prove that Newton's third law of motion is obeyed in the gravitational interaction between two masses.

2.2. Derive Eq. (2.2) geometrically from the first principles.

2.3. Verify Eq. (2.8).

2.4. Verify Eqs. (2.9) to (2.12).

2.5. Prove the reflective property of the ellipse (see Appendix A.3).

2.6. Verify Eq. (2.44).

2.7. Show that the magnitude of the gravitational field due to a point mass is isotropic.

2.8. Show that the magnitude of the gravitational field due to a point mass is spherically symmetric.

2.9. Show that the gravitational force due to a point mass is an irrotational vector, i.e., $\vec{\nabla} \times \vec{f} = \vec{0}$.

2.10. Show that the gravitational force due to a point mass is a solenoidal vector, i.e., $\vec{\nabla} \cdot \vec{f} = 0$.

2.11. Show that the gravitational force due to a point mass is a true vector, i.e., $\vec{f}(-\vec{r}) = -\vec{f}(\vec{r})$.

2.12. Show that the gravitational potential energy at a point given by the two definitions in Sec. 2.6 are equivalent.

2.13. Show that gravitational force satisfies all three criteria of a conservative force.

2.14. Verify Eq. (2.94).

Average and Extremum Values of Variables

3.1 Concerning the Distance in Kepler's Third Law

The original version of Kepler's third law stated that the square of the period of a planet was proportional to the cube of its "mean distance" from the Sun. We now know that this "mean distance" must refer to the semi-major axis and nothing else. That is, the period of the planet depends only on the major axis and not on the minor axis. Therefore, the "mean distance' must not contain the eccentricity in it. In order to satisfy this condition, it is often said that the "mean distance" refers to the "mean of the greatest and the smallest distances of the planet from the Sun", i.e., the "mean of the aphelion and perihelion distances". However, we know that there exist three common means between the two: the "arithmetic mean", the "geometric mean" and the "harmonic mean".

Given two quantities p and q, the three means are defined respectively as follows:

$$(AM) = \frac{p+q}{2}, \qquad (3.1)$$

$$(GM) = \sqrt{pq}, \qquad (3.2)$$

and

$$(HM) = \frac{2pq}{p+q}. \tag{3.3}$$

Two other means also exist: the "quadratic mean" (also called the "root mean square") and the "harmonic root mean" [cf. Iles and Wilson (1980) and Tan (1982)]. Between p and q, they are defined as

$$(QM) = \sqrt{\frac{p^2 + q^2}{2}}, \tag{3.4}$$

and

$$(HRM) = \sqrt{\frac{2p^2q^2}{p^2 + q^2}}. \tag{3.5}$$

It can be verified that

$$(AM)(HM) = (GM)^2 = (QM)(HRM). \tag{3.6}$$

The geometric mean between two quantities is also equal to the geometric mean between their arithmetic and harmonic means and further is equal to the geometric mean between their quadratic and harmonic root means!

For the planetary orbit, the perihelion and aphelion distances are given by

$$r_P = a(1 - e), \tag{1.6}$$

and

$$r_A = a(1 + e). \tag{1.7}$$

The five means between r_P and r_A are readily calculated from Eqs. (3.1)–(3.5) and entered in Table 3.1. The three most common means, viz., the arithmetic mean, the geometric mean and the harmonic mean are respectively equal to the semi-major axis, the semi-minor axis and the semi-latus rectum of the orbital ellipse. Evidently, it is the arithmetic mean which is the appropriate mean in Kepler's original version of the third law.

Table 3.1. Common means between perihelion and aphelion distances.

Mean	Mean value	Familiar distance
Quadratic mean	$a\sqrt{1+e^2}$	
Arithmetic mean	a	Semi major axis
Geometric mean	$a\sqrt{1-e^2}$	Semi minor axis
Harmonic mean	$a(1-e^2)$	Semi latus rectum
Harmonic root mean	$\dfrac{a(1-e^2)}{\sqrt{1+e^2}}$	

3.2 Average Values of the Radial Distance

Since the radial distance of the planet from the Sun varies continuously between its extreme values r_P and r_A, it is more appropriate to consider its average over an entire orbit rather than the means of the extremum values. Once again, three kinds of averages come to mind: (i) the average of r over the angular coordinate θ; (ii) the average of r over time t; and (iii) the average of r over the path s.

First, the average of r over θ can be expressed as

$$\langle r \rangle_\theta = \frac{\int_0^{2\pi} r\, d\theta}{\int_0^{2\pi} d\theta} = \frac{2\int_0^{\pi} r\, d\theta}{2\pi}. \tag{3.7}$$

Utilizing the equation of the ellipse (1.5), we have

$$\langle r \rangle_\theta = \frac{p}{\pi} \int\limits_0^\pi \frac{d\theta}{1 + e\cos\theta}. \tag{3.8}$$

The simple-looking integral in Eq. (3.8) is actually not an easy matter to deal with. Using standard integral tables, we get

$$\int \frac{d\theta}{1 + e\cos\theta} = \frac{2}{\sqrt{1-e^2}} \tan^{-1} \sqrt{\frac{1-e}{1+e}} \tan\frac{\theta}{2}. \tag{3.9}$$

The right-hand side of Eq. (3.9) can be simplified by the following trigonometric identity

$$2 \tan^{-1} \sqrt{\frac{1-e}{1+e}} \tan \frac{\theta}{2} = \cos^{-1} \frac{e + \cos \theta}{1 + e \cos \theta}, \qquad (3.10)$$

whence by putting the limits in Eq. (3.8), we get

$$\langle r \rangle_\theta = a\sqrt{1 - e^2} = b. \qquad (3.11)$$

The average distance over the polar angle is equal to the semi-minor axis. This is not the desired distance for Kepler's third law.

Next we discuss the average distance of the planet over time

$$\langle r \rangle_t = \frac{\int_0^P r \, dt}{\int_0^P dt} = \frac{1}{P} \int_0^P r \, dt. \qquad (3.12)$$

By a change of variable, one can write [cf. Stein (1977)]

$$\langle r \rangle_t = \frac{1}{P} \int_0^{2\pi} r \frac{dt}{d\theta} d\theta. \qquad (3.13)$$

Writing $\dot{A}$ for the constant areal velocity, we have

$$P = \frac{\pi ab}{\dot{A}}, \qquad (3.14)$$

and

$$\frac{dt}{d\theta} = \frac{r^2}{2\dot{A}} = \frac{mr^2}{l}. \qquad (3.15)$$

Thus, we get [cf. Stein (1977)]

$$\langle r \rangle_t = \frac{1}{2\pi ab} \int_0^{2\pi} r^3 \, d\theta = \frac{p^3}{2\pi ab} \int_0^{2\pi} \frac{d\theta}{(1 + e \cos \theta)^3}. \qquad (3.16)$$

The integral in Eq. (3.16) can be found by successive substitutions, but that exercise is even more cumbersome than the one in the previous example. It can also be performed by using a software such as *Mathematica*. We only give the final result here:

$$\langle r \rangle_t = a \left(1 + \frac{e^2}{2} \right). \tag{3.17}$$

Clearly, this too, is not the desired length in Kepler's third law.

Finally, for the average distance over the perimeter of the ellipse, we have

$$\langle r \rangle_s = \frac{\int_0^L r\,ds}{\int_0^L ds} = \frac{1}{L} \int_0^L r\,ds, \tag{3.18}$$

where L is the length of the perimeter of the ellipse. The latter is given by an elliptical integral for which there is no exact solution. Approximate expressions for L found in the literature include

$$L \approx \pi\sqrt{2(a^2 + b^2)}, \tag{3.19}$$

and

$$L \approx \pi \left[3(a + b) - \sqrt{(a + 3b)(3a + b)} \right]. \tag{3.20}$$

The last one is due to the famous Indian mathematician Ramanujan [cf. Weisstein (2003)]. In any case, the exact solution for this last average has to be handled by other methods. For all three averages, there is lack of simple elegant solutions. In the following sections, we discuss alternative methods for calculating the average values of the radial distance.

If one defines the most probable distance of the planet as the radial distance where the planet spends the most time for the same angular displacement, then that is the distance where $dt/d\theta$ is maximum, which according to Eq. (3.15), is where r is maximum. Thus, the most probable distance is equal to the aphelion distance of the planet, or $a(1 + e)$.

3.3 True and Eccentric Anomalies

In polar coordinates *apropos* Keplerian ellipse, the angular coordinate is measured at the focus. It is sometimes more convenient to use the angle measured at the center of the ellipse, called the *eccentric angle* (cf. Appendix A.11). This angle is obtained by dropping the perpendicular QS from the planet at Q to the major axis and extending it up to R on the auxiliary circle (Fig. 3.1). Consistent with Kepler's usage, the polar angle θ is called the *true anomaly*, whereas the eccentric angle E is called the *eccentric anomaly* [cf. Bate *et al.* (1971)].

Since any chord perpendicular to the major axis is shortened from the auxiliary circle to the ellipse by the ratio of $\sqrt{1 - e^2}$, we have from Fig. 3.1,

$$r \cos \theta = a(\cos E - e), \qquad (3.21)$$

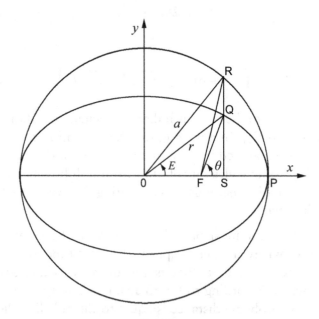

Fig. 3.1

and

$$r \sin \theta = a \sin E \sqrt{1 - e^2}. \tag{3.22}$$

By squaring and adding Eqs. (3.21) and (3.22), we get r as a function of E,

$$r = a(1 - e \cos E). \tag{3.23}$$

The true anomaly can be expressed in terms of the eccentric by eliminating r between Eqs. (3.21) and (3.22) [cf. Van de Kamp (1960)],

$$\cos \theta = \frac{\cos E - e}{1 - e \cos E}, \tag{3.24}$$

and

$$\sin \theta = \frac{\sqrt{1 - e^2} \sin E}{1 - e \cos E}. \tag{3.25}$$

Algebraic manipulation yields the inverse relations

$$\cos E = \frac{e + \cos \theta}{1 + e \cos \theta}, \tag{3.26}$$

and

$$\sin E = \frac{\sqrt{1 - e^2} \sin \theta}{1 + e \cos \theta}. \tag{3.27}$$

One can construct the following ratio:

$$\frac{1 - \cos \theta}{1 + \cos \theta} = \frac{1 + e}{1 - e} \frac{1 - \cos E}{1 + \cos E}, \tag{3.28}$$

which leads to

$$\tan \frac{\theta}{2} = \sqrt{\frac{1 + e}{1 - e}} \tan \frac{E}{2}. \tag{3.29}$$

The above equation is widely-used to calculate the true anomaly from the eccentric anomaly.

3.4 Kepler's Equation, Mean Anomaly and Mean Motion

The *equation of time* can be obtained with reference to Fig. 3.1 where Q is the location of the planet, S the foot of the perpendicular on the major axis, R is the intersection of SQ with the auxiliary circle, and the rest have their usual meaning. If t_0 is the time of the perihelion passage of the planet, then by Kepler's second law,

$$\frac{t - t_0}{P} = \frac{areaFPQ}{\pi ab}. \tag{3.30}$$

Remembering that every ordinate of the ellipse is shortened from the corresponding ordinate on the auxiliary circle by a factor of $\sqrt{1 - e^2}$, we have

$$areaFPQ = \sqrt{1 - e^2}\, areaFPR. \tag{3.31}$$

Now

$$areaFPR = areaOPR - areaOSR + areaFSR, \tag{3.32}$$

$$areaOPR = \frac{1}{2}Ea^2, \tag{3.33}$$

$$areaOSR = \frac{1}{2}(a\cos E)(a\sin E) = \frac{1}{2}a^2\cos E\sin E, \tag{3.34}$$

and

$$areaFSR = \frac{1}{2}(a\cos E - ae)(a\sin E)$$

$$= \frac{1}{2}a^2\cos E\sin E - \frac{1}{2}a^2 e\sin E. \tag{3.35}$$

From Eqs. (3.31)–(3.35), we have

$$areaFPQ = \sqrt{1 - e^2}\frac{a^2}{2}(E - e\sin E) = \frac{ab}{2}(E - e\sin E). \tag{3.36}$$

Hence Eq. (3.30) reduces to

$$E - e\sin E = \frac{2\pi(t - t_0)}{P}. \tag{3.37}$$

From the expression for the period derived in Chap. 2,

$$P = 2\pi\sqrt{\frac{a^3}{GM}}, \qquad (1.70)$$

whence

$$E - e\sin E = \sqrt{\frac{GM}{a^3}}(t - t_0). \qquad (3.38)$$

The right-hand side of Eq. (3.38) is also called the *mean anomaly* M. Written in terms of M, Eq. (3.38) is called *Kepler's equation*,

$$E - e\sin E = M. \qquad (3.39)$$

Finally, the *mean motion* is defined by

$$n = \frac{2\pi}{P} = \sqrt{\frac{GM_m}{a^3}}. \qquad (3.40)$$

The mean motion represents the average angular speed of the planet over a complete revolution. Kepler's equation, written in full, is now

$$E - e\sin E = n(t - t_0) = M. \qquad (3.41)$$

Note that in Eq. (3.40), the mass of the attracting body has been re-written as M_m in order to distguish it from the newly-defined mean anomaly.

3.5 Average Values of Radial Distance using Eccentric Anomaly

Calculating the average distance of the planet is facilitated by a change of variable to the eccentric anomaly E. For the average distance of r over θ, we have [cf. Van de Kamp (1964)]

$$\langle r \rangle_\theta = \frac{\int_0^\pi r d\theta}{\int_0^\pi d\theta} = \frac{\int_0^\pi r \frac{d\theta}{dE} dE}{\pi}. \tag{3.42}$$

In order to find $d\theta/dE$, we first differentiate Eq. (3.37) to obtain

$$(1 - e \cos E) dE = \frac{2\pi}{P} dt. \tag{3.43}$$

Substituting from Eq. (3.23),

$$\frac{dE}{dt} = \frac{2\pi a}{rP}. \tag{3.44}$$

Also, from Kepler's second law,

$$\frac{1}{2} r^2 \frac{d\theta}{dt} = \frac{\pi a b}{P}. \tag{3.45}$$

From Eqs. (3.44) and (3.45), one gets

$$\frac{d\theta}{dE} = \frac{b}{r}, \tag{3.46}$$

whence from Eq. (3.42),

$$\langle r \rangle_\theta = \frac{\int_0^\pi b dE}{\pi} = b = a\sqrt{1 - e^2}. \tag{3.47}$$

Next, for the average of the radial distance r over t, we have [cf. Van de Kamp (1964)]

$$\langle r \rangle_t = \frac{\int_0^{P/2} r dt}{\int_0^{P/2} dt} = \frac{\int_0^\pi r \frac{dt}{dE} dE}{\frac{P}{2}}. \tag{3.48}$$

Making substitutions for dt/dE from Eq. (3.44) and r from Eq. (3.23), we get

$$\langle r \rangle_t = \frac{\int_0^\pi a^2 (1 - e \cos E)^2 dE}{\pi a} = \frac{a}{\pi} \int\limits_0^\pi (1 - 2e \cos E + e^2 \cos^2 E) dE. \tag{3.49}$$

Upon carrying out the three integrals, we get

$$\langle r \rangle_t = a \left(1 + \frac{e^2}{2} \right). \tag{3.50}$$

Finally, for the average of the radial distance r over s, we have [cf. Tan and Chameides (1981)]

$$\langle r \rangle_s = \frac{\int_0^L r \, ds}{\int_0^L ds} = \frac{\int_0^\pi r \frac{ds}{dE} dE}{\int_0^\pi \frac{ds}{dE} dE}. \tag{3.51}$$

If x and y are the abscissa and ordinate of the planet, respectively, referring to the center of the ellipse as origin (Fig. 3.1), then

$$x = a \cos E, \tag{3.52}$$

and

$$y = a\sqrt{1 - e^2} \sin E. \tag{3.53}$$

Differentiating Eqs. (3.52) and (3.53) with respect to E, squaring and adding, we get

$$\frac{ds}{dE} = a\sqrt{1 - e^2 \cos^2 E}. \tag{3.54}$$

Substitution of ds/dE from Eq. (3.54) and r from Eq. (3.23) in Eq. (3.51) gives

$$\langle r \rangle_s = a - ae \frac{\int_0^\pi \sqrt{1 - e^2 \cos^2 E} \cos E \, dE}{\int_0^\pi \sqrt{1 - e^2 \cos^2 E} \, dE}. \tag{3.55}$$

The denominator of the second term on the right-hand side of Eq. (3.55) is just the perimeter of the ellipse. The numerator can be evaluated readily and it vanishes upon putting the limits. Thus, we are left with

$$\langle r \rangle_s = a. \tag{3.56}$$

One can also calculate the average value of r over the eccentric anomaly itself (Prussing, 1977):

$$\langle r \rangle_E = \frac{\int_0^\pi r\, dE}{\int_0^\pi dE} = \frac{\int_0^\pi a(1 - e\cos E)dE}{\pi} = a. \tag{3.57}$$

Likewise, over the mean anomaly,

$$\langle r \rangle_M = \frac{\int_0^\pi r\, dM}{\int_0^\pi dM} = \frac{\int_0^\pi a(1 - e\cos E)dM}{\pi}. \tag{3.58}$$

From Kepler's equation,

$$dM = (1 - e\cos E)dE. \tag{3.59}$$

Hence

$$\langle r \rangle_M = \frac{a}{\pi} \int_0^\pi (1 - e\cos E)^2 dE = a\left(1 + \frac{e^2}{2}\right). \tag{3.60}$$

Table 3.2 summarizes the average values of the radial distance over the different variables. Two of the averages, over the perimeter and the eccentric anomaly, produce the appropriate distance for Kepler's third law.

Table 3.2. Average values of the radial distance.

Variable	Average value	Familiar distance
True anomaly	$a\sqrt{1 - e^2}$	Semi-minor axis
Time	$a\left(1 + \frac{e^2}{2}\right)$	
Perimeter	a	Semi-major axis
Eccentric anomaly	a	Semi-major axis
Mean anomaly	$a\left(1 + \frac{e^2}{2}\right)$	

3.6 Locations of Extrema of Dynamical Variables

In this section, we examine the locations of the planet given by the angular coordinate θ where the dynamical variables attain their maximum and minimum values. Firstly, it is easy to see from the polar equation of the ellipse that r is minimum at the perihelion ($\theta = 0$) and maximum at the aphelion ($\theta = \pi$). Secondly, Kepler's second law states that the areal velocity is constant everywhere on the orbit. Thirdly, the *angular velocity* can be expressed in terms of the constant angular momentum:

$$\omega = \frac{d\theta}{dt} = \frac{l}{mr^2}. \tag{3.61}$$

It is obviously maximum at the perihelion ($\theta = 0$) and minimum at the aphelion ($\theta = \pi$).

The *radial velocity* of the planet can be obtained via a chain rule:

$$v_r = \frac{dr}{dt} = \frac{dr}{d\theta}\frac{d\theta}{dt} = \frac{pe\sin\theta}{(1 + e\cos\theta)^2}\frac{l}{mr^2} = \frac{l}{mp}\sin\theta. \tag{3.62}$$

Clearly, v_r is extremum where $\sin\theta$ is: it is maximum at $\theta = \pi/2$ and minimum at $\theta = 3\pi/2$. The planet ascends and descends the fastest at the ends of the latus rectum through the focus.

For the *radial acceleration*, we consider its absolute value only. From Newton's law of gravitation, we have

$$|a_r| = \left|\frac{d^2r}{dt^2}\right| = \frac{GM}{r^2}. \tag{3.63}$$

Clearly, the magnitude of the radial acceleration is the greatest at the perihelion ($\theta = 0$) and the least at the aphelion ($\theta = \pi$). The locations of the extrema of the centripetal acceleration are also found at the same places.

Finally, the *angular acceleration* of the planet can be expressed by a double chain rule (Tan, 1988),

$$\alpha = \frac{d\omega}{dt} = \frac{d\omega}{dr}\frac{dr}{d\theta}\frac{d\theta}{dt}. \tag{3.64}$$

Gathering the various factors from the equation of the ellipse and Eq. (3.61), one gets

$$\alpha = -\frac{4l^2 e}{m^2 p^4}(1 + e\cos\theta)^3 \sin\theta. \tag{3.65}$$

In order to find the locations of the extrema of α, we differentiate Eq. (3.65) with respect to θ and set $d\alpha/d\theta = 0$, whence

$$4e\cos^2\theta + \cos\theta - 3e = 0, \tag{3.66}$$

and

$$\cos\theta = \frac{-1 \pm \sqrt{1 + 48e^2}}{8e}. \tag{3.67}$$

Since $\cos\theta > 0$ for $0 < \theta < \pi/2$, we retain the upper sign only (the lower sign corresponds to an extraneous solution), getting

$$\theta = \cos^{-1}\frac{\sqrt{1 + 48e^2} - 1}{8e}, \tag{3.68}$$

or

$$\theta = 2\pi - \cos^{-1}\frac{\sqrt{1 + 48e^2} - 1}{8e}. \tag{3.69}$$

An inspection of Eq. (3.65) indicates that α is actually negative for $0 < \theta < \pi$. Thus the solutions given by Eqs. (3.68) and (3.69) correspond to maximum and minimum angular accelerations, respectively. Equation (3.65) also reveals that the angular acceleration is zero at $\theta = 0$ (perihelion) and $\theta = \pi$ (aphelion). To illustrate, for $e = 0.1, 0.5$ and 0.9, the angular accelerations are minima at $74.24°, 49.35°$ and $42.42°$, respectively. The corresponding angles for angular acceleration maxima are compliments of these angles from 2π.

Table 3.3. Angular positions of extrema of dynamical variables.

Variable	θ where maximum	θ where minimum
r	0	π
$v_r = \frac{dr}{dt}$	$\pi/2$	$3\pi/2$
$\lvert a_r \rvert = \left\lvert \frac{d^2 r}{dt^2} \right\rvert$	0	π
$\frac{1}{2} r^2 \frac{d\theta}{dt}$	Same everywhere	
$v_\theta = r\frac{d\theta}{dt}$	0	π
$v = \sqrt{v_r^2 + v_\theta^2}$	0	π
$\omega = \frac{d\theta}{dt}$	0	π
$\alpha = \frac{d^2\theta}{dt^2}$	$\cos^{-1} \frac{\sqrt{1+48e^2}-1}{8e}$	$2\pi - \cos^{-1} \frac{\sqrt{1+48e^2}-1}{8e}$

Table 3.3 summarizes the extremum locations of the planet pertaining to the various dynamical variables.

3.7 Centripetal Acceleration in Planetary Motion

The centripetal acceleration of a planet at any point in its orbit is given by [cf. Eq. (2.2)]

$$a_c = \frac{v^2}{\rho}, \qquad (3.70)$$

where v is the velocity and ρ is the radius of curvature at that point. In terms of our usual symbols, we have from Eqs. (2.42) and (2.8)

$$v^2 = \frac{l^2}{m^2 p^2} \left(1 + e^2 + 2e\cos\theta \right), \qquad (3.71)$$

and

$$\rho = \frac{p \left(1 + e^2 + 2e\cos\theta \right)^{3/2}}{(1 + e\cos\theta)^3}, \qquad (3.72)$$

whence

$$\frac{v^2}{\rho} = \frac{l^2}{m^2 p^3} f(\theta), \tag{3.73}$$

where

$$f(\theta) = \frac{(1 + e\cos\theta)^3}{\sqrt{1 + e^2 + 2e\cos\theta}}. \tag{3.74}$$

The locations of centripetal acceleration extremum are obtained by differentiating $f(\theta)$ with respect to θ and setting the derivative equal to zero (Tan, 1991). We have

$$\frac{df}{d\theta} = -\frac{e\sin\theta(1 + e\cos\theta)^2(2 + 3e^2 + 5e\cos\theta)}{(1 + e^2 + 2e\cos\theta)^{3/2}}. \tag{3.75}$$

Since the factor $(1 + e\cos\theta)^2$ and the denominator of the right-hand side of Eq. (3.75) are always positive, the conditions of the extremum centripetal acceleration are determined by the vanishing of the factors $\sin\theta$ and $(2 + 3e^2 + 5e\cos\theta)$. Now $\sin\theta$ is always zero at $\theta = 0$ and $\theta = \pi$ whereas $(2 + 3e^2 + 5e\cos\theta)$ vanishes only for $e \geq 2/3$ at

$$\theta = \cos^{-1}\frac{-(2 + 3e^2)}{5e}, \tag{3.76}$$

and

$$\theta = 2\pi - \cos^{-1}\frac{-(2 + 3e^2)}{5e}. \tag{3.77}$$

Hence we have two distinct extremum cases: (i) for $e < 2/3$, the extrema are found at $\theta = 0$ (perihelion) and $\theta = \pi$ (aphelion); and (ii) for $e > 2/3$, the same are found at $\theta = 0, \theta = \pi$, and the locations given by Eqs. (3.76) and (3.77). An inspection of the signs of $df/d\theta$ on either sides of the extrema indicates that in the first case ($e < 2/3$), the centripetal acceleration is maximum at the perihelion and minimum at the aphelion. In the more interesting second case ($e > 2/3$), the centripetal acceleration is still maximum at the perihelion but is minimum at the locations (3.76)

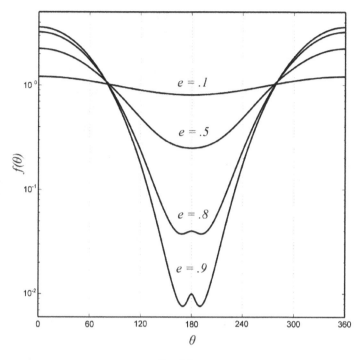

Fig. 3.2

and (3.77). In this case, the aphelion corresponds to a secondary maximum. As examples, for $e = 0.7$, 0.8 and 0.9, the secondary maximum straddles the minima which are located at $(172.5°, 187.5°)$, $(168.5°, 191.5°)$ and $(169.9°, 190.1°)$, respectively.

Figure 3.2 is a logarithmic plot of $f(\theta)$ versus θ for selected values of the eccentricity e. It shows that the variation of $f(\theta)$ increases dramatically with e. For $e = 0.9$, for instance, the maximum value of $f(\theta)$ and hence that of the centripetal acceleration is over 475 times its minimum value. Figure 3.2 also indicates that the perihelion ($\theta = 0$) is always the location of maximum $f(\theta)$. The aphelion ($\theta = \pi$), on the other hand, is the location of minimum $f(\theta)$ for $e < 2/3$ only and corresponds to a secondary maximum for $e > 2/3$. Only in the limiting case of a circular orbit ($e = 0$) is the centripetal acceleration constant throughout the orbit.

3.8 The Jerk Vector in Planetary Motion

The jerk vector is seldom discussed in the literature [cf. Sandin (1990)]. As defined in Eq. (1.20), it is the first, second or third derivative of the acceleration, velocity or position vectors with respect to time, respectively. Being a higher derivative, it is usually a null vector or a small vector. For example, the jerk vector is zero for uniformly accelerated motion in a straight line and in projectile motion without air resistance under gravity. For uniform circular motion or simple harmonic motion, the jerk vector is anti-parallel to the velocity vector [cf. Sandin (1990) and Tan (1992b)].

For planetary motion, we begin with the position vector

$$\vec{r} = r\hat{r} = \frac{p}{1 + e\cos\theta}\hat{r}. \tag{3.78}$$

Differentiating Eq. (3.78) twice with respect to t by using the chain rule and Eqs. (1.21) and (1.22), we obtain

$$\vec{v} = \frac{le\sin\theta}{mp}\hat{r} + \frac{l(1 + e\cos\theta)}{mp}\hat{\theta}, \tag{3.79}$$

and

$$\vec{a} = -\frac{l^2(1 + e\cos\theta)^2}{m^2p^3}\hat{r}. \tag{3.80}$$

Equation (3.80) signifies that the acceleration vector has no transverse component. This is, of course, consistent with the central force problem. A further differentiation furnishes the jerk vector

$$\vec{j} = j_r\hat{r} + j_\theta\hat{\theta}, \tag{3.81}$$

where

$$j_r = \frac{l^3}{m^3p^5}2e\sin\theta(1 + e\cos\theta)^3, \tag{3.82}$$

and

$$j_\theta = -\frac{l^3}{m^3 p^5}(1 + e\cos\theta)^4. \tag{3.83}$$

The magnitude of the jerk vector follows from Eqs. (3.82) and (3.83):

$$j = \sqrt{j_r^2 + j_\theta^2}$$
$$= \frac{l^3}{m^3 p^5}(1 + e\cos\theta)^3 \sqrt{1 + 4e^2 + 2e\cos\theta - 3e^2\cos^2\theta}. \tag{3.84}$$

For the special case of the circular orbit, we obtain by putting $e = 0$, $p = a$ and $l = mva$:

$$\vec{j} = -\frac{v^3}{a^2}\hat{\theta}, \tag{3.85}$$

a result found in Sandin (1990).

From the planet's point of view, it is instructive to calculate the components of the jerk vector in the forward and sideward directions. Let j_s be the forward jerk and j_t be the sideward jerk reckoned positive if to the right. Then from Fig. 3.3,

$$j_s = j_r \sin\phi + j_\theta \cos\phi, \tag{3.86}$$

and

$$j_t = j_r \cos\phi - j_\theta \sin\phi, \tag{3.87}$$

where ϕ is the angle between the radius vector and the normal to the path. Also, we see that

$$v_\theta = v\cos\phi, \tag{3.88}$$

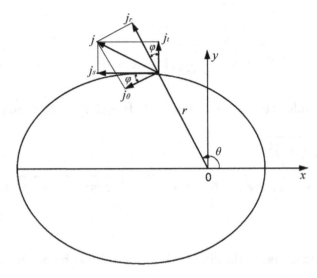

Fig. 3.3

and

$$v_r = v \sin \phi. \tag{3.89}$$

Utilizing Eqs. (2.39), (2.40) and (2.41), Eqs. (3.88) and (3.89) give

$$\cos \phi = \frac{1 + e \cos \theta}{\sqrt{1 + e^2 + 2e \cos \theta}}, \tag{3.90}$$

and

$$\sin \phi = \frac{e \sin \theta}{\sqrt{1 + e^2 + 2e \cos \theta}}. \tag{3.91}$$

By substituting Eqs. (3.82), (3.82), (3.90) and (3.91) in Eqs. (3.86) and (3.87), we finally arrive at expressions for the forward and sideward jerks as functions of θ only (Tan, 1992b):

$$j_s = -\frac{l^3}{m^3 p^5} \frac{(1 + e \cos \theta)^3 (3e^2 \cos^2 \theta + 2e \cos \theta - 2e^2 + 1)}{\sqrt{1 + e^2 + 2e \cos \theta}}, \tag{3.92}$$

and

$$j_t = \frac{l^3}{m^3 p^5} \frac{3e \sin\theta (1 + e\cos\theta)^4}{\sqrt{1 + e^2 + 2e\cos\theta}}. \tag{3.93}$$

Figure 3.4 is a plot of the jerk vector and its components as functions of θ for an orbit having an eccentricity of $e = 0.3$ [from Tan (1992b)]. Even for this relatively small eccentricity, the magnitudes of j and its components show strong dependences on θ. For example, the magnitudes of j, j_θ and j_s at the perihelion are about an order of magnitude greater than those at the aphelion. j_s, like j_θ, is always negative over the entire orbit which means that the jerk vector is always backwards in planetary motion. j_r and j_t, on the other hand, change signs as the planet crosses the line of apsides and are positive for the ascending phase of the planet, i.e., for increasing r. The jerk would turn from the right to the left as the

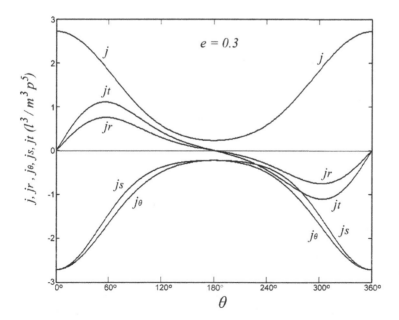

Fig. 3.4

planet crosses the aphelion and turn back to the right as it crosses the perihelion.

The extrema of j, j_θ and j_s are found at the apsidal points, but those of j_r and j_t lie somewhere near $60°$ and $300°$. The latter locations are found by differentiating Eq. (3.82) with respect to θ and setting the derivative equal to zero. We have

$$\frac{dj_r}{d\theta} = \frac{l^3}{m^3 p^5}(1 + e\cos\theta)^2(4e\cos^2\theta + \cos\theta - 3e). \qquad (3.94)$$

Since $(1 + e\cos\theta)^2$ is always positive, the extrema of j_r are given by the equation

$$4e\cos^2\theta + \cos\theta - 3e = 0, \qquad (3.95)$$

i.e.,

$$\cos\theta = \frac{-1 \pm \sqrt{1 + 48e^2}}{8e}. \qquad (3.96)$$

Retaining the positive sign only (the negative sign gives imaginary solutions), we have

$$\theta = \cos^{-1}\frac{\sqrt{1 + 48e^2} - 1}{8e}, \qquad (3.97)$$

or

$$\theta = 2\pi - \cos^{-1}\frac{\sqrt{1 + 48e^2} - 1}{8e}. \qquad (3.98)$$

The first solution (3.97) gives the maximum of j_r whereas the second solution (3.98) gives its minimum. Note that they are the same locations where the angular acceleration has its extrema [Eqs. (3.68) and (3.69)]. For the orbit with $e = 0.3$, the two locations are at $\theta = 57.02°$ and $\theta = 302.98°$.

Differentiating Eq. (3.86) once more with respect to θ, one obtains

$$\frac{d^2 j_r}{d\theta^2} = -\frac{l^3}{m^3 p^5} \sin\theta(1 + e\cos\theta)(16e^2 \cos^2\theta + 11\cos\theta - 6e^2 + 1).$$

$$(3.99)$$

By inspection, this derivative is zero at $\theta = 0$ and $\theta = \pi$, even though the first derivative is not zero at these locations. The perihelion and aphelion are thus inflexion points for j_r.

3.9 Polar Coordinates in Reference to the Empty Focus

It is sometimes advantageous to use polar coordinates (r', θ') in reference to the empty focus F' (Fig. 3.5). We shall utilize such coordinates to obtain a few results. First, we re-calculate the average of the radial distance over the length of the perimeter L. From symmetry, we have [cf. Stein (1977)]

$$\int_0^L r\,ds = \int_0^L r'\,ds.$$

$$(3.100)$$

By virtue of the property that the sum of the distances from any point on the ellipse to the two foci is constant and equal to the major axis, we get

$$\int_0^L r\,ds = \frac{1}{2}\left[\int_0^L r\,ds + \int_0^L r'\,ds\right] = \frac{1}{2}\int_0^L (r + r')\,ds = \frac{1}{2}\int_0^L 2a\,ds = aL.$$

$$(3.101)$$

Hence, from Eq. (3.18),

$$\langle r\rangle_s = \frac{\int_0^L r\,ds}{L} = a,$$

$$(3.102)$$

a result obtained earlier.

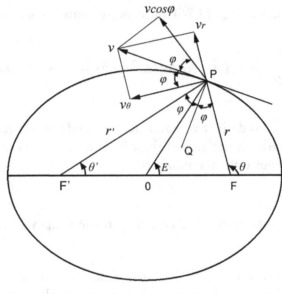

Fig. 3.5

We next calculate the angular velocity of the planet in reference to the empty focus. In Fig. 3.5, PQ is drawn perpendicular to the velocity vector which is tangential to the orbit. By the well-known reflective property of the ellipse, $\angle FPQ = \angle QPF' = \phi$. v_r and v_θ are the radial and transverse components of the velocity $\vec{v}$ of the planet and $v \cos \phi$ is the transverse component of the velocity in reference to F'. It is easy to see that v_θ and $v \cos \phi$ make the same angle ϕ with the velocity vector and are therefore equal in magnitude: $v_\theta = v \cos \phi$. The angular velocity in reference to the empty focus is thus

$$\omega' = \frac{d\theta'}{dt} = \frac{v \cos \phi}{r'} = \frac{v_\theta}{r'}. \qquad (3.103)$$

In terms of the constant angular momentum l, we have

$$v_\theta = r\frac{d\theta}{dt} = \frac{l}{mr}. \qquad (3.104)$$

Hence

$$\omega' = \frac{l}{mrr'}. \qquad (3.105)$$

In terms of the eccentric anomaly, we had

$$r = a(1 - e \cos E). \tag{3.23}$$

By another well-known property of the ellipse,

$$r + r' = 2a, \tag{3.106}$$

whence

$$r' = a(1 + e \cos E). \tag{3.107}$$

Substituting Eqs. (3.23) and (3.107) in Eq. (3.105) and expanding using the binomial series, we get

$$\omega' = \frac{l}{ma^2(1 - e \cos E)(1 + e \cos E)} = \frac{l}{ma^2}\left(1 + e^2 \cos^2 E + \cdots\right). \tag{3.108}$$

If one retains terms up to e^2 only, then

$$\omega' \approx \frac{l}{ma^2}\left(1 + e^2 \cos^2 E\right). \tag{3.109}$$

In general, $\cos^2 E$ has the same values at four locations of the orbit: $E = E, E = \pi{-}E, E = \pi + E$ and $E = 2\pi -E$ and so does ω'. These points are symmetrically situated with respect to the major and minor axes. For orbits having small eccentricities, terms containing e^2 and higher powers of e could be neglected when we have

$$\omega' \approx \frac{l}{ma^2} = const. \tag{3.110}$$

Thus, we have the following theorem.

Theorem 3.1. *For small eccentricities, the angular velocity in reference to the empty focus is a constant. Equation (3.110) constitutes a historical alternative to Kepler's second law, which was widely used to track the planets for half a century before it was abandoned [for a popular discussion, see Cohen (1981)]. For all planets in the solar system except Mercury (which was difficult to observe) and Pluto (which was yet undiscovered), the maximum error encountered in Eq. (3.110) was less than one percent.*

The maximum errors for Mercury (e = 0.206) and Pluto (e = 0.249)
are four percent and six percent respectively, which occur at the ends
of the major axis (E = 0 or π).

3.10 Estimating the Eccentricity of Earth's Orbit

To continue the discussion in Sec. 1.13, one can also estimate the
eccentricity of the Earth's orbit from the dates of the equinoxes.
Let T denote the time interval of passage of the Earth from the
autumnal equinox E' (September 22/23) to the vernal equinox E
(March 20); T' the co-interval from E to E'; and Y the planet's year.
Then $T = 178.5$ days; $Y = 365.25$ days; and $T' = 186.75$ days.
The ratio $R = T/Y = 714/1461$. The small but significant difference
between T and T' can be exploited to determine the eccentricity of
the Earth's orbit (Tan, 2004–2005). We can do so by calculating the
intervals and comparing them with the observed values. There are
three options: one, compare T and T'; two, compare T' and Y; and
three, compare T and Y. The third option happens to be the most
convenient.

We can begin by separating the variables in Eq. (1.36),

$$\int dt = \frac{m}{l} \int r^2 d\theta. \tag{3.111}$$

Substituting from Eq. (1.5),

$$\int dt = \frac{mp^2}{l} \int \frac{d\theta}{(1 + e\cos\theta)^2}. \tag{3.112}$$

Ignoring the constant mp^2/l, we can write

$$I = \int \frac{d\theta}{(1 + e\cos\theta)^2}. \tag{3.113}$$

The integral I can be evaluated by using standard integral tables, giving

$$I = -\frac{e \sin \theta}{(1 - e^2)(1 + e \cos \theta)^2} + \frac{2}{(1 - e^2)^{3/2}} \tan^{-1} \sqrt{\frac{1 - e}{1 + e}} \tan \frac{\theta}{2}.$$
(3.114)

We can now compute the relative magnitudes of T and Y. We have

$$Y \propto 2[I]_0^\pi = \frac{2\pi}{(1 - e^2)^{3/2}}.$$
(3.115)

For the interval T, we can write

$$T \propto [I]_{3\pi/2-\alpha}^{3\pi/2} + 2[I]_0^{\pi/2} - [I]_{\pi/2-\alpha}^{\pi/2}.$$
(3.116)

It is apparent that the first and third terms on the right-hand side of the proportionality constant are numerically very nearly equal, with the first fractionally greater than the third. They, therefore, nearly cancel each other out, giving

$$T \propto 2[I]_0^{\pi/2} \approx -\frac{2e}{1 - e^2} + \frac{4}{(1 - e^2)^{3/2}} \tan^{-1} \sqrt{\frac{1 - e}{1 + e}}.$$
(3.117)

One can simplify the last term in Eq. (3.117) by a trigonomotric identity to obtain [cf. Aravind (1987)]

$$T \propto -\frac{2e}{1 - e^2} + \frac{2 \cos^{-1} e}{(1 - e^2)^{3/2}}.$$
(3.118)

Setting $T/Y = R$ from Eqs. (3.118) and (3.115), we arrive at

$$-e\sqrt{1 - e^2} + \cos^{-1} e = \frac{714\pi}{1461}.$$
(3.119)

This is a transcendental equation of the form

$$f(e) = g(e), \qquad\qquad (3.120)$$

where

$$f(e) = \cos^{-1} e, \qquad\qquad (3.121)$$

and

$$g(e) = \frac{714\pi}{1461} + e\sqrt{1 - e^2}. \qquad\qquad (3.122)$$

The transcendental equation (3.120) is customarily solved by the graphical method (Fig. 3.6). The intersection of the two graphs $f(e)$ and $g(e)$ gives the solution $e = 0.0177$. The result compares

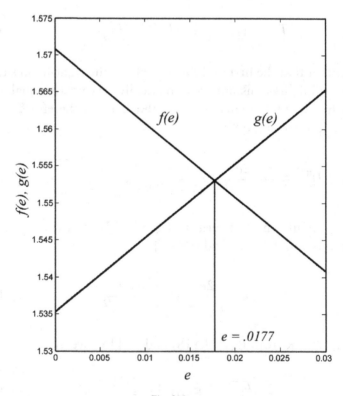

Fig. 3.6

very favorably with the actual value of 0.017 and only marginally overestimates it.

Alternatively, one can approximate the first two terms on the left-hand side of Eq. (3.119) by series expansions [cf. Aravind (1987)] giving

$$e\sqrt{1 - e^2} = e(1 - e^2)^{1/2} = e\left(1 - \frac{e^2}{2} - \frac{e^4}{8} - \cdots\right), \qquad (3.123)$$

and

$$\cos^{-1} e = \frac{\pi}{2} - \sin^{-1} e = \frac{\pi}{2} - e - \frac{e^3}{6} - \cdots. \qquad (3.124)$$

Substituting in Eq. (3.119) and retaining terms up to the first order in e, we arrive at

$$e \approx \frac{33\pi}{5844} \approx 0.018. \qquad (3.125)$$

The value of e is still quite close to the actual value of 0.017.

For orbits having small eccentricities, we can proceed with yet another approach to the integral (3.113). We can use an approximation to the integral, but keep the limits intact. This method is due to D. Roaf [cf. Tan (2004–2005)]. Upon series expansion of the denominator and retention of terms up to the first order in e only,

$$I \approx \int (1 - 2e\cos\theta)d\theta = \theta - 2e\sin\theta. \qquad (3.126)$$

In this case,

$$Y \propto 2[I]_0^{\pi} = 2\pi, \qquad (3.127)$$

and

$$T \propto [I]_{3\pi/2 - \alpha}^{\pi/2 - \alpha} = \pi - 4e\cos\alpha. \qquad (3.128)$$

Once again, by setting $T/Y = R$ and solving, we arrive at (Tan 2004–2005)

$$e = \frac{\pi}{4\cos\alpha}(1 - 2R). \qquad (3.129)$$

Putting $R = 714/1461$ and $\alpha = 13.3°$, we get $e \approx 0.018$. This value is, once again, quite close to that of the actual value of the eccentricity.

3.11 Estimating the Distance of a Heavenly Body

The distances of heavenly bodies from the Earth are determined by the *Parallax Method*. If pictures of a heavenly body are taken from two distant locations on the Earth and compared, the displacement of the body on the background of the distant star field (see Sec. 10.3 for Celestial Sphere) furnishes a determination of the distance of that body from the Earth. If x is the linear distance between the two locations on the Earth and d is the distance of the heavenly body from the Earth, then the angular displacement of the body on the Celestial Sphere is given by

$$\theta = \frac{x}{d}. \qquad (3.130)$$

From the observed angular displacement θ and linear distance x, the distance of the heavenly body is readily found. The nearer the heavenly body to the Earth, the greater is its displacement.

Exercises

3.1. Verify Eq. (3.6).
3.2. Derive Eq. (3.9).
3.3. Prove the relation (3.10).
3.4. Derive Eq. (3.17).

3.5. Verify Eqs. (3.24) to (3.27).

3.6. Derive Eq. (3.29).

3.7. Verify Eqs. (3.31) to (3.35).

3.8. Derive the inverse relation of Eq. (3.29), i.e.,

$$\tan \frac{E}{2} = \sqrt{\frac{1-e}{1+e}} \tan \frac{\theta}{2}.$$

3.9. Show that $\left\langle \frac{1}{r} \right\rangle_t = \frac{1}{a}$ and $\left\langle \frac{1}{r} \right\rangle_\theta = \frac{1}{p} = \frac{1}{a(1-e^2)}$.

3.10. Calculate $\left\langle \frac{1}{r^3} \right\rangle_t$ and $\left\langle \frac{1}{r^4} \right\rangle_t$.

3.11. Using the Virial Theorem, show that $\left\langle v^2 \right\rangle_t = \frac{GM}{a}$ (see, Bucher and Siemens, 1998).

3.12. Verify Eq. (3.75).

3.13. Verify Eqs. (3.86) and (3.87).

3.14. Verify Eqs. (3.90) and (3.91).

3.15. Verify Eqs. (3.92) and (3.93).

3.16. Verify Eq. (3.94).

3.17. Verify Eq. (3.99).

3.18. Verify Eq. (3.114).

3.19. Verify Eq. (3.118).

The Central Force Problem

4.1 Central Forces

A *central force* is one which is always directed towards a central point in space. It is radial in direction and has no transverse component. In spherical coordinates with the force center as the origin, this force can be written as

$$\vec{f} = f(r)\hat{r} = g(r)\vec{r}. \tag{4.1}$$

Such a force is irrotational and usually conservative. Thus it can be expressed as the negative gradient of a scalar potential energy

$$\vec{f} = -\vec{\nabla}V(r). \tag{4.2}$$

Table 4.1 lists some of the common central forces observed in nature, together with their respective potentials and their dependence on the radial coordinate.

Table 4.1.   Common central forces.

Example	Radial dependence	
	Force	Potential
Simple Harmonic Oscillator	r	r^2
Surface Charge Distribution	const.	r
Line Charge Distribution	$\dfrac{1}{r}$	$\log_e r$
Gravitational Force/Electric Force	$\dfrac{1}{r^2}$	$\dfrac{1}{r}$
Electric Dipole	$\dfrac{1}{r^3}$	$\dfrac{1}{r^2}$
Electric Quadrupole	$\dfrac{1}{r^4}$	$\dfrac{1}{r^3}$
Electric Octupole	$\dfrac{1}{r^5}$	$\dfrac{1}{r^4}$
Van der Waal's Force	$\dfrac{1}{r^7}$	$\dfrac{1}{r^6}$
Yukawa Potential		$\dfrac{e^{-\alpha r}}{r}$
General Theory of Relativity	$\dfrac{1}{r^2}$ plus $\dfrac{1}{r^4}$	

4.2 General Dynamics of Angular Motion

The *moment of momentum* or *angular momentum* of a particle about the origin is defined as

$$\vec{l} = \vec{r} \times \vec{p}, \tag{4.3}$$

where the notations have their general meaning. The *moment of force* or *torque* about the same origin is similarly defined as

$$\vec{\tau} = \vec{r} \times \vec{f}. \tag{4.4}$$

Vector differentiation of Eq. (4.3) gives

$$\frac{d\vec{l}}{dt} = \frac{d\vec{r}}{dt} \times \vec{p} + \vec{r} \times \frac{d\vec{p}}{dt} = \vec{\tau}. \tag{4.5}$$

Equation (4.5) is the analog of Newton's second law in angular motion.

For a central force given by Eq. (4.1), we have

$$\vec{\tau} = \vec{r} \times g(r)\vec{r} = \vec{0}. \tag{4.6}$$

Hence, by Eq. (4.5), the angular momentum in central force motion is always constant. As a consequence, both the position vector and the momentum vector (and thus the velocity vector) must remain in the same plane according to Eq. (4.3), since any change in direction of either vector will cause a change in direction of the angular momentum vector and violate its constancy. Kepler's law of areas is thus generally valid for all central force problems.

4.3 The Planetary Problem in Polar Coordinates

The formal derivations of Kepler's laws of planetary motion from Newton's law of gravitation are now laid out in polar coordinates. As demanded, the Sun and the planet are considered point masses, and the mass of the planet m is assumed to be negligible compared with that of the Sun M so that the Sun maintains its position at the origin. First, the absence of a transverse force on the planet ensures the validity of Kepler's second law and the conservation of the orbital angular momentum of the planet,

$$l = mr^2 \frac{d\theta}{dt} = const. \tag{4.7}$$

The radial component of the force equation is obtained from Eqs. (1.31) and (2.47):

$$m \frac{d^2 r}{dt^2} - mr \left(\frac{d\theta}{dt} \right)^2 = -\frac{GMm}{r^2}. \tag{4.8}$$

To obtain the equation of the orbit, one must eliminate time t between Eqs. (4.7) and (4.8). The customary and traditional treatment of the problem calls for the substitution of the variable

$$u = \frac{1}{r}. \tag{4.9}$$

By successive chain rules, we have

$$\frac{dr}{dt} = \frac{dr}{du}\frac{du}{d\theta}\frac{d\theta}{dt} = -\frac{l}{m}\frac{du}{d\theta}. \tag{4.10}$$

By a further chain rule,

$$\frac{d^2r}{dt^2} = \frac{d}{dt}\left(\frac{dr}{dt}\right) = \frac{d}{d\theta}\left(\frac{dr}{dt}\right)\frac{d\theta}{dt} = -\frac{l^2}{m^2}u^2\frac{d^2u}{d\theta^2}. \tag{4.11}$$

Furthermore,

$$r\left(\frac{d\theta}{dt}\right)^2 = \frac{l^2}{m^2}u^3. \tag{4.12}$$

Equation (4.8) now assumes the form

$$\frac{d^2u}{d\theta^2} + u - \frac{GMm^2}{l^2} = 0. \tag{4.13}$$

A second substitution

$$x = u - \frac{GMm^2}{l^2}, \tag{4.14}$$

reduces the radial equation of motion into a simpler form

$$\frac{d^2x}{d\theta^2} + x = 0. \tag{4.15}$$

Equation (4.15) is the equation of a *simple harmonic motion* with an *angular frequency* or *angular velocity* $\omega = 1$. Its solutions are periodic functions of sines and cosines of θ. With a proper choice of boundary conditions, one can write

$$x = A\cos\theta, \tag{4.16}$$

where A is the amplitude as yet undetermined. Substituting back for x and u from Eqs. (4.14) and (4.9), we arrive at the polar equation of a conic

$$r = \frac{\frac{l^2}{GMm^2}}{1 + \frac{Al^2}{GMm^2} \cos \theta}. \tag{4.17}$$

Equation (4.17) is recognized as the polar equation of a conic with a semi-latus rectum

$$p = \frac{l^2}{GMm^2}, \tag{4.18}$$

and eccentricity

$$e = \frac{Al^2}{GMm^2}. \tag{4.19}$$

Equation (4.17) accounts for Kepler's first law (closed elliptical orbits, including circular orbits), but also allows open parabolic (e.g. escape trajectories) and hyperbolic trajectories (e.g. planetary fly-bys and scattering trajectories). However, because of the amplitude factor A, the eccentricity in Eq. (4.19) remains as yet undetermined. Figure 4.1 shows the family of conic sections for allowable orbits and trajectories under the inverse square force law having the same perihelion distances.

   We can complete the discussion in this section with the derivation of Kepler's third law. From Kepler's second law (1.35) and the conservation of the angular momentum of the planet, we have

$$\frac{dA}{dt} = \frac{1}{2}r^2\frac{d\theta}{dt} = \frac{l}{2m}. \tag{4.20}$$

Separating the variables and integrating over one complete revolution, we get

$$\int_0^P dt = \frac{2m}{l} \int_0^A dA, \tag{4.21}$$

or

$$P = \frac{2m}{l}\pi ab = \frac{2\pi m}{l}a^2\sqrt{1 - e^2}. \tag{4.22}$$

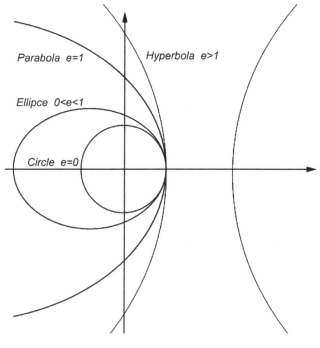

Fig. 4.1

Eliminating the eccentricity from Eq. (1.4), we arrive at

$$P = \frac{2\pi m \sqrt{p}}{l} a^{\frac{3}{2}}, \qquad (4.23)$$

which is Kepler's harmonic law [Eq. (1.1)].

4.4 Alternative Derivation of the Equation of Orbits

An alternative method of integration for the orbit furnishes the eccentricity in terms of the total energy of the planet [cf. Goldstein (1980)]. We begin with the total energy of the planet:

$$\frac{1}{2}m \left(\frac{dr}{dt}\right)^2 + \frac{1}{2}mr^2 \left(\frac{d\theta}{dt}\right)^2 - \frac{GMm}{r} = E. \qquad (4.24)$$

Writing the second term on the left-hand side in terms of l from Eq. (4.7), and rearranging, we get

$$\frac{dr}{dt} = \sqrt{\frac{2}{m}\left(E + \frac{GMm}{r} - \frac{l^2}{2mr^2}\right)}. \qquad (4.25)$$

We can now use the chain rule to eliminate time. By virtue of Eq. (4.7), the left-hand side of Eq. (4.25) is converted into

$$\frac{dr}{dt} = \frac{dr}{d\theta}\frac{d\theta}{dt} = \frac{l}{mr^2}\frac{dr}{d\theta}, \qquad (4.26)$$

whence, Eq. (4.25) becomes

$$\frac{dr}{d\theta} = \frac{mr^2}{l}\sqrt{\frac{2}{m}\left(E + \frac{GMm}{r} - \frac{l^2}{2mr^2}\right)}. \qquad (4.27)$$

By separating the variables and integrating, we obtain

$$\theta - \theta_0 = \int \frac{dr}{r\sqrt{\frac{2mEr^2}{l^2} + \frac{2GMm^2r}{l^2} - 1}}. \qquad (4.28)$$

The right-hand side of Eq. (4.28) can be evaluated using standard integration tables, giving:

$$\theta - \theta_0 = \sin^{-1}\frac{\frac{GMm^2}{l^2} - \frac{1}{r}}{\sqrt{\frac{G^2M^2m^4}{l^2} + \frac{2mE}{l^2}}}. \qquad (4.29)$$

Setting $\theta_0 = 3\pi/2$, and manipulating, we arrive at [cf. Norwood (1979)]

$$r = \frac{\frac{l^2}{GMm^2}}{1 + \sqrt{1 + \frac{2l^2E}{G^2M^2m^3}}\cos\theta}. \qquad (4.30)$$

Once again, we have the standard equation of a conic in polar coordinates having semi-latus rectum

$$p = \frac{l^2}{GMm^2}, \qquad (4.31)$$

and eccentricity

$$e = \sqrt{1 + \frac{2l^2 E}{G^2 M^2 m^3}}. \qquad (4.32)$$

Equation (4.31) is identical to Eq. (4.18). But unlike in Eq. (4.19), the eccentricity of the orbit is now determined in terms of the known quantities in Eq. (4.32).

The total energy of the planet can be expressed in terms of the angular momentum and eccentricity of the orbit,

$$E = \frac{G^2 M^2 m^3}{2l^2}(e^2 - 1). \qquad (4.33)$$

For a circular orbit ($e = 0$), we get, by substituting from Eq. (2.75),

$$E_c = -\frac{G^2 M^2 m^3}{2l^2} = -\frac{GMm}{2a}, \qquad (4.34)$$

which is an expression obtained earlier [see Eq. (2.87)]. For a parabolic trajectory ($e = 1$), $E = 0$. For a hyperbolic trajectory ($e > 1$), $E > 0$. Note also that for the hyperbolic orbit, a is negative [cf. Bate *et al.* (1971)]. The nature of the various conical orbits and trajectories are summarized in Table 4.2, with the relevant parameters.

Table 4.2.   Inverse square law orbits/trajectories.

Nature of orbit/ trajectory	Whether open or closed	Eccentricity	Semi-major axis	Total energy
Circular	Closed	$e = 0$	$a = p > 0$	$E_c < 0$
Elliptical (including circular)	Closed	$0 \leq e < 1$	$a > p > 0$	$E_c < E_e < 0$
Parabolic	Open	$e = 1$	$0 < p < a = \infty$	$E_p = 0$
Hyperbolic	Open	$e > 1$	$a < 0 < p$	$E_h > 0$

4.5 Lagrangian and Hamiltonian Formulations of Classical Mechanics

Constraints are conditions imposed on a system which limit the degrees of freedom of the system. *Holonomic constraints* are those which are expressible in the form of algebraic equations. *Lagrange's equations* for a conservative holonomic system with the forces of constraint performing no work, are written as

$$\frac{d}{dt}\left(\frac{\partial L}{\partial \dot{q}_i}\right) - \frac{\partial L}{\partial \dot{q}_i} = 0, \quad i = 1, 2, \ldots, n, \tag{4.35}$$

where

$$L = T - V, \tag{4.36}$$

is the *Lagrangian*, q_i's are the independent *generalized coordinates* (linear coordinates and angular coordinates), n is the number of independent coordinates, and

$$\dot{q}_i = \frac{dq_i}{dt}, \tag{4.37}$$

is the *generalized velocity* (linear velocity or angular velocity). Lagrange's equations yield the same equations of motion as the force and torque equations. They, therefore, do not furnish anything new other than a different perspective.

In the Lagrangian formulation, the *generalized momentum* (linear momentum or angular momentum) p_i corresponding to the generalized velocity $\dot{q}_i$ is defined by

$$p_i = \frac{\partial L}{\partial \dot{q}_i}. \tag{4.38}$$

If a generalized coordinate q_i does not appear in the Lagrangian explicitly, then by Eqs. (4.35) and (4.38)

$$\frac{d}{dt}(p_i) = 0, \tag{4.39}$$

or, the corresponding generalized momentum is a constant of motion. That particular generalized coordinate is then called a *cyclic coordinate* or an *ignorable coordinate* [cf. Whittaker (1961)].

The *Hamiltonian* is defined by

$$H = \sum_{i=1}^{n} p_i \dot{q}_i - L, \qquad (4.40)$$

where p_i and q_i are now called *conjugate variables*. The Hamiltonian is equal to the total energy of the system if (i) the kinetic energy is a homogeneous quadratic function of the generalized velocities; and (ii) if the potential energy is not a function of velocity (i.e. the forces are conservative).

Whereas the Lagrangian is a function of the generalized coordinates and velocities, the Hamiltonian is conventionally expressed as a function of the generalized coordinates and generalized momenta by transforming the generalized velocities into generalized momenta:

$$L = L(q_i, \dot{q}_i, t), \qquad (4.41)$$

but

$$H = H(q_i, p_i, t). \qquad (4.42)$$

Hamilton's equations of motion for conservative holonomic systems are written as

$$\dot{q}_i = \frac{\partial H}{\partial p_i}, i = 1, 2, \ldots, n, \qquad (4.43)$$

and

$$\dot{p}_i = -\frac{\partial H}{\partial q_i}, i = 1, 2, \ldots, n. \qquad (4.44)$$

There are thus twice as many Hamilton's equations as there are Lagrange's equations. However, only one set of equations (4.44) is equivalent to Lagrange's equations, the other set of equations (4.43) merely gives the definitions of the generalized momenta.

Hamiltonian mechanics provides yet another perspective of classical mechanics, but like Lagrangian mechanics, it does not provide any additional result.

It is easy to verify that Hamilton's equations (4.43) are equivalent to the definitions of the conjugate momenta, while Eq. (4.44) are equivalent to Lagrange's equations of motion.

4.6 The Planetary Problem in Lagrangian and Hamiltonian Formulations

The Lagrangian for the planetary problem is given, in polar coordinates, by

$$L = T - V = \frac{1}{2}m\left(\frac{dr}{dt}\right)^2 + \frac{1}{2}mr^2\left(\frac{d\theta}{dt}\right)^2 + \frac{GMm}{r}, \qquad (4.45)$$

where the symbols carry their usual meaning. The θ-equation gives the familiar conservation of the angular momentum of the planet

$$p_\theta = l = mr^2\frac{d\theta}{dt} = const. \qquad (4.46)$$

Clearly, θ is an ignorable coordinate. The r-equation gives

$$m\frac{d^2r}{dt^2} - mr\left(\frac{d\theta}{dt}\right)^2 + \frac{GMm}{r^2} = 0. \qquad (4.47)$$

Equations (4.46) and (4.47) are identical to Eqs. (4.7) and (4.8), respectively. Technically, they are coupled equations in θ and r. However, the substitution of Eqs. (4.46) into (4.47) yields

$$m\frac{d^2r}{dt^2} - \frac{l^2}{mr^3} + \frac{GMm}{r^2} = 0. \qquad (4.48)$$

Thus, the planetary problem is reduced to a one-dimensional problem in r only. In a coordinate system rotating with the planet, the motion is entirely in the radial direction. This illustrates the significance of the ignorable coordinate.

In the Hamiltonian formulation, the conjugate momenta are given by

$$p_r = \frac{\partial L}{\partial \dot{r}} = m\dot{r}, \qquad (4.49)$$

and

$$p_\theta = \frac{\partial L}{\partial \dot{\theta}} = mr^2\dot{\theta}. \qquad (4.50)$$

The Hamiltonian works out to be

$$H = p_r\dot{r} + p_\theta\dot{\theta} - L = T - V = E, \qquad (4.51)$$

as expected. Expressed in terms of the conjugate momenta

$$H = \frac{p_r^2}{2m} + \frac{p_\theta^2}{2mr^2} - \frac{GMm}{r}. \qquad (4.52)$$

4.7 Variational Principles in Classical Mechanics

Variational principle deals with the study of the extremum (usually minimum) of integrals. Given the independent variable x, the dependent variable y and its derivative $y' = dy/dx$, if the integral of a function $f(y, y', x)$ is minimum between two points on its path, then that of a neighboring path is the same to a first order. Mathematically, this is written as

$$d \int_{x_1}^{x_2} f(y, y', x)dx = 0. \qquad (4.53)$$

The condition of minimum yields the *Euler–Lagrange Equation*:

$$\frac{d}{dx}\left(\frac{\partial f}{\partial y'}\right) - \frac{\partial f}{\partial y} = 0. \qquad (4.54)$$

In Mechanics, the variational principle for a conservative holonomic system is found:

$$d \int_{t_1}^{t_2} L(q_i, \dot{q}_i, t) dt = 0. \qquad (4.55)$$

This is called **Hamilton's principle**. Lagrange's equation (4.35) immediately follow from the Euler–Lagrange condition. Conversely, Hamilton's principle can also be obtained from Lagrange's equations [cf. Greenwood (1977)].

An alternative form of Hamilton's principle is referred to as the **Principle of least action**:

$$d \int_{t_1}^{t_2} 2T dt = 0. \qquad (4.56)$$

Furthermore, if the kinetic energy is a homogeneous quadratic function of the generalized velocities, then one obtains *Jacobi's form of the principle of least action* [cf. Greenwood (1977)]:

$$d \int_{s_1}^{s_2} \sqrt{2T} ds = 0, \qquad (4.57)$$

where ds is the arc length

$$ds = \sqrt{dr^2 + r^2 d\theta^2}. \qquad (4.58)$$

4.8 The Planetary Problem from the Variational Principle

For the planetary problem, one can express T in terms of the total energy E, which is a constant of motion

$$T = E + \frac{GMm}{r}. \qquad (4.59)$$

Also, the arc length can be expressed in terms of an unspecified independent variable u [cf. Osgood (1965)]:

$$ds = \sqrt{r'^2 + r^2\theta'^2}\,du, \qquad (4.60)$$

where the prime indicates differentiation with respect to u. Jacobi's form of the principle of least action then assumes the form

$$d\int_1^2 \sqrt{E + \frac{GMm}{r}}\sqrt{r'^2 + r^2\theta'^2}\,du = 0. \qquad (4.61)$$

Here

$$f(r, r'; \theta, \theta') = \sqrt{E + \frac{GMm}{r}}\sqrt{r'^2 + r^2\theta'^2}. \qquad (4.62)$$

Since f does not contain θ explicitly, it is an ignorable coordinate. The Euler–Lagrange equation furnishes a constant of integration

$$\frac{\partial f}{\partial \theta'} = \sqrt{E + \frac{GMm}{r}}\frac{r^2\theta'}{\sqrt{r'^2 + r^2\theta'^2}} = c. \qquad (4.63)$$

Taking the independent variable as θ [cf. Osgood (1965)], we have

$$\sqrt{E + \frac{GMm}{r}}\frac{r^2}{\sqrt{\left(\frac{dr}{d\theta}\right)^2 + r^2}} = c. \qquad (4.64)$$

Squaring and re-arranging,

$$\frac{d\theta}{dr} = \frac{c}{r\sqrt{Er^2 + GMmr - c^2}}. \qquad (4.65)$$

By separating the variables and integrating, we get

$$\theta = c\int \frac{dr}{r\sqrt{Er^2 + GMmr - c^2}} = \sin^{-1}\frac{GMmr - 2c^2}{r\sqrt{G^2M^2m^2 + 4Ec^2}}. \qquad (4.66)$$

Finally, this is re-cast into a form of an equation of the conic (cf. Sec. 5.3)

$$r = \frac{\frac{2c^2}{GMm}}{1 - \sqrt{1 + \frac{4Ec^2}{G^2 M^2 m^2}} \sin \theta}. \qquad (4.67)$$

4.9 The Effective Potential Energy

By integrating the force equation (4.48) over r, one obtains the energy equation

$$\int m \frac{d^2 r}{dt^2} dr - \int \frac{l^2}{mr^3} dr + \int \frac{GMm}{r^2} dr = E. \qquad (4.68)$$

The first term on the left can be integrated with a change in variable

$$\int m \frac{d^2 r}{dt^2} dt = \int m \frac{d^2 r}{dt^2} \frac{dr}{dt} dt = \int \frac{d}{dt} \left[\frac{1}{2} m \left(\frac{dr}{dt} \right)^2 \right] dt. \qquad (4.69)$$

Equation (4.68) then integrates out to

$$\frac{1}{2} m \left(\frac{dr}{dt} \right)^2 + \frac{l^2}{2mr^2} - \frac{GMm}{r} = E. \qquad (4.70)$$

In the rotating coordinate system, the equation of energy is written thus

$$T + V' = E, \qquad (4.71)$$

where V' is the *effective potential energy*

$$V' = \frac{l^2}{2mr^2} - \frac{GMm}{r}. \qquad (4.72)$$

The first terms in V' is the *centrifugal potential energy* which invariably appears in a rotating coordinate system, whereas the second term is the familiar *gravitational potential energy*.

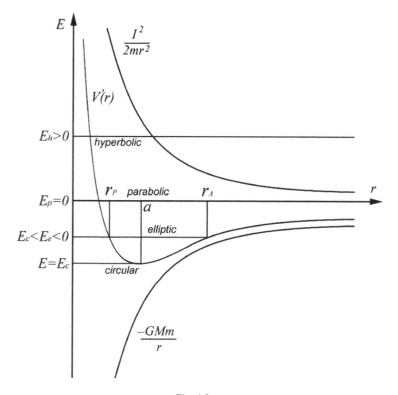

Fig. 4.2

Figure 4.2 depicts the effective potential energy as well as its constituent terms as functions of the radial coordinate. Also shown in the figure are the total energy levels of the four conic section orbits or trajectories from Table 4.2. The perihelion and aphelion distances of the elliptical orbit as well as the radius of the circular orbit are evident in the figure.

In order to examine the minimum of V', we differentiate Eq. (4.72) twice to obtain

$$\frac{dV'}{dr} = -\frac{l^2}{mr^3} + \frac{GMm}{r^2}, \tag{4.73}$$

and

$$\frac{d^2V'}{dr^2} = \frac{3l^2}{mr^4} - \frac{2GMm}{r^3}. \tag{4.74}$$

The condition of the extremum $dV'/dr = 0$ gives the radius of the circular orbit

$$r = a = \frac{l^2}{GMm^2}, \tag{4.75}$$

and the energy of the circular orbit

$$V'(a) = E_c = -\frac{GMm}{2a}. \tag{4.76}$$

One can verify that the condition of minimum of V' is satisfied:

$$\frac{d^2V'}{dr^2} = \frac{GMm}{r^3} > 0. \tag{4.77}$$

4.10 Determination of the Force Law from the Equation of Orbit

The normal procedure for deriving the equation of orbit from the force law consists of eliminating the time between the equations of motion and solving the resulting differential equation. The process of integration is often complicated when the attractive force is not inverse square in nature. An alternative method of study is to prescribe the equation of the orbit from which to determine the force law. This inverse approach happens to be far simpler than the direct problem, since it involves differentiation only instead of complicated integration. Indeed, historically, the latter problem was called the "direct problem" whereas the former problem was known as the "inverse problem" [cf. Weinstock (1992)].

For a general central force problem, the force equation is written as [vide Eq. (4.8)]

$$f(r) = m\frac{d^2r}{dt^2} - mr\left(\frac{d\theta}{dt}\right)^2. \tag{4.78}$$

Eliminating t from Eqs. (4.7), (4.9) and (4.11), we have the equation of the central force in terms of the orbital parameters

$$f(r) = f\left(\frac{1}{u}\right) = -\frac{l^2u^2}{m}\left(\frac{d^2u}{d\theta^2} + u\right), \tag{4.79}$$

with

$$u = \frac{1}{r}. \tag{4.9}$$

Given the equation of the orbit in plane polar coordinates, one can easily find the r-dependence of the central force.

As an exercise, consider the general equation of a conic [Eq. (1.5)]. We have

$$u = \frac{1}{p} + \frac{e}{p}\cos\theta. \tag{4.80}$$

By differentiating twice with respect to θ,

$$\frac{d^2u}{d\theta^2} = -\frac{e}{p}\cos\theta, \tag{4.81}$$

whence from Eq. (4.79)

$$f(r) = -\frac{l^2}{mp}\frac{1}{r^2} \propto \frac{1}{r^2}, \tag{4.82}$$

which is the inverse square law of gravitation.

Problems of this nature have formed interesting topics in many textbooks [e.g. MacMillan (1927), Lamb (1961) and Fowles

Table 4.3. Orbits/trajectories and central forces.

Orbit/trajectory	Polar equation of curve	Nature of curve	Radial dependence of force
Conic Sections	$r = \dfrac{p}{1 + e\cos\theta}$	Trigonometric	$\dfrac{1}{r^2}$
Hyperbolic Spiral or Reciprocal Spiral	$r = \dfrac{a}{\theta}$	Transcendental	$\dfrac{1}{r^3}$
Logarithmic Spiral or Equiangular Spiral	$r = e^{a\theta}$	Transcendental	$\dfrac{1}{r^3}$
Archimedean Spiral	$r = a\theta$	Transcendental	$\dfrac{1}{r^3}$ plus $\dfrac{1}{r^5}$
Parabolic Spiral	$r = a\theta^2$	Transcendental	$\dfrac{1}{r^3}$ plus $\dfrac{1}{r^4}$
Fermat's Spiral	$r = a\sqrt{\theta}$	Transcendental	$\dfrac{1}{r^3}$ plus $\dfrac{1}{r^7}$
Circle	$r = 2a\cos\theta$	Trigonometric	$\dfrac{1}{r^5}$
Cardioid	$r = a(1 + \cos\theta)$	Trigonometric	$\dfrac{1}{r^4}$
Lemniscate of Bernoulli	$r^2 = a^2\cos 2\theta$	Trigonometric	$\dfrac{1}{r^7}$

(1962)]. Table 4.3 is a collection of various orbits and trajectories and the underlying central forces which produce them. The polar equations of the plane curves are given, together with the nature of the curves. Several interesting observations can be made from the table. First, conic sections (including elliptical orbits, parabolic escape trajectories and hyperbolic fly-byes) are only allowed under the inverse square law of gravitation. Second, various kinds of spiral trajectories are allowed under the inverse-cube law of force. Third, some spirals include a term having higher inverse r-dependence, but all spirals necessarily contain the inverse-cube term. Fourth, the spiral trajectories are called *transcendental curves*, by which are meant curves which intersect a radial line at an infinite number of points [cf. Weisstein (2003)]. Fifth, there exists a circular orbit under the inverse fifth power on one point which is located at the attracting center. The latter also holds true for the cardioid and the lemniscate.

4.11 A General Orbit Equation and the Force Law

It is interesting to note that most of the plane curves of the earlier section belong to the general family of curves given by

$$r^n = a^n \cos n\theta, \tag{4.83}$$

where n is a rational number [cf. Lawrence (1972)]. We can carry out the usual procedure of determining the power law. We have

$$u = \frac{1}{r} = \frac{1}{a}(\cos n\theta)^{-\frac{1}{n}}, \tag{4.84}$$

$$\frac{du}{d\theta} = \frac{1}{a}(\sin n\theta)(\cos n\theta)^{-\frac{n+1}{n}}, \tag{4.85}$$

$$\frac{d^2u}{d\theta^2} = \frac{n}{a}(\cos n\theta)^{-\frac{1}{n}} + \frac{n+1}{a}(\cos n\theta)^{-\frac{2n+1}{n}}(\sin n\theta)^2, \tag{4.86}$$

and

$$\frac{d^2u}{d\theta^2} + u = \frac{n+1}{a}(\cos n\theta)^{-\frac{2n+1}{n}}, \tag{4.87}$$

whence from Eq. (4.79),

$$f(r) = -\frac{l^2(n+1)a^{2n}}{m}\frac{1}{r^{2n+3}} \propto \frac{1}{r^{2n+3}}. \tag{4.88}$$

Equation (4.83) represents a family of curves, some of which are listed in Table 4.4. The corresponding power law of the force is also shown in the table, together with examples of such forces in nature. The orbital curves (except the spirals) are plotted in Fig. 4.4 [from Tan (1981)]. The numbers correspond to the inverse power of the radial distance.

The curve with $n = -2$ is a rectangular hyperbola whose asymptotes are the lines having slopes ± 1 [cf. Lawrence (1972)]. The curve with $n = -3/2$ is also a hyperbola, but having asymptotes making angles of $\pm\pi/3$ with the x-axis. It corresponds to the attraction of an infinite surface distribution of matter (or charge). $n = 1$ is the

Table 4.4.   Nature of orbit and force law.

Order of curve (n)	The curve	Inverse power of $r(2n + 3)$	Example in nature
−2	Hyperbola	−1	
−3/2	Hyperbola	0	Infinite surface distribution
−1	Straight line	1	Infinite line distribution
−1/2	Parabola	2	Point distribution
0	Spirals	3	Electric dipole
1/2	Cardioid	4	Electric quadrupole
1	Circle	5	Electric octupole
2	Lemniscate	7	Van der Waal's force

case of an infinite line distribution of matter or charge. In the figure, this line passes through the origin and is perpendicular to the plane of the paper. $n = -1/2$ (the parabola) is an example of the inverse square law, to which we shall return later. The case of $n = 0$ belongs to the spirals, which have been dealt with at length by Lamb (1961). The cases of $n = 1$ and $n = 2$ can be found in MacMillan (1927) and also in Lamb (1961). The latter case corresponds to the weak Van der Waals' forces between two neutral atoms or molecules [cf. Born (1969)].

The cardioid presents a case for which the force varies as the inverse fourth power of the distance. It may be recalled that in Einstein's **General Theroy of Relativity**, the gravitational field may be thought to be inverse square with a small inverse fourth power term [cf. Eddington (1963)]. The inverse square term, of course, gives the Newtonian orbit, while the inverse fourth power is responsible for the **precession of perihelion**. Geometrically, we might think of the perihelic precession to be the result of the perturbation of a weak cardioid on the dominant elliptical orbit.

The list of the curves is, be no means exhaustive. There are curves higher up and lower down in the ladder and also in between the entries (note that n is any rational number and does not have to be an integer). However, a few interesting observations can be made from Table 4.4 and Fig. 4.3. It is easy to see that curves higher than the parabolic pass through the attracting center. Hence, the formation of planets is not possible in such worlds, only collisions are. The curves

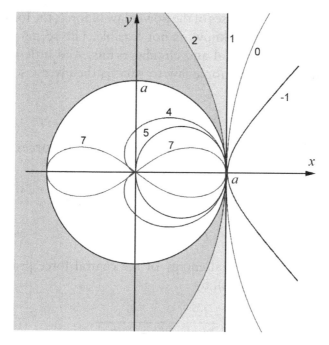

Fig. 4.3

on the other side of the scale are all hyperbolic in nature, with similar consequences. Therefore only scattering is possible in such worlds.

One should notice that in the case of the inverse square law of force ($n = -1/2$), only the parabola is represented in the general equation (4.68). Nevertheless, all four conics, which are legitimate trajectories under the inverse square law of force, are shown in Fig. 4.3. Among them, the parabolic and hyperbolic trajectories are unbounded and represent scattering. Only the elliptical and circular trajectories constitute stable planetary orbits. The shaded areas of Fig. 4.3 show the domain of the inverse square force trajectories. The lightly-shaded area is where the hyperbolic trajectories ($e > 1$) lie. The darkly-shaded area, on the other hand, is the domain of the elliptical orbits ($e < 1$). It is here that the stable planetary orbits can be found. This includes the circle ($e = 0$), but excludes the parabola ($e = 1$).

To sum up, in most cases of the power law of force, the formation of a solar system as we know, is not feasible. Life seems possible only in the stable elliptical and circular orbits. And little wonder, mother nature chose the force law to vary as the inverse square of the distance!

4.12 Stability of Circular Orbits under Central Forces

If we assume an attractive central force of the form

$$f(r) = -\frac{k}{r^n}, \tag{4.89}$$

then the effective potential energy of the central force problem is given, in lieu of Eq. (4.72), by

$$V' = \frac{l^2}{2mr^2} - \frac{k}{(n-1)r^{n-1}}. \tag{4.90}$$

In that case

$$\frac{dV'}{dr} = -\frac{l^2}{mr^3} - \frac{k}{r^n}, \tag{4.91}$$

and

$$\frac{d^2V'}{dr^2} = \frac{3l^2}{mr^4} - \frac{nk}{r^{n+1}}. \tag{4.92}$$

The conditions of stable equilibrium are given by

$$\frac{dV'}{dr} = 0, \tag{4.93}$$

and

$$\frac{d^2V'}{dr^2} > 0. \tag{4.94}$$

Equations (4.91) and (4.93) give

$$r^{n-3} = \frac{mk}{l^2}, \tag{4.95}$$

whence, from Eqs. (4.92), (4.94) and (4.95),

$$\frac{l^2}{m}(3 - n) > 0. \tag{4.96}$$

Thus, the condition for a stable circular orbit is that n must be less than three [cf. Marion and Thornton (1995)].

An alternative approach to this problem is given as follows [cf. Norwood (1979)]. For the central force (4.89), the equation of motion in the radial direction is given by

$$m\frac{d^2r}{dt^2} - \frac{l^2}{mr^3} = -\frac{k}{r^n}. \tag{4.97}$$

For a circular orbit of radius a, the centripetal force is

$$\frac{l^2}{ma^3} = \frac{k}{a^n}. \tag{4.98}$$

Consider a small radial displacement x from the circular path

$$r = a + x, \tag{4.99}$$

where $a \gg x$. Then

$$\frac{d^2r}{dt^2} = \ddot{x}. \tag{4.100}$$

The equation of motion (4.97) assumes the form

$$m\ddot{x} - \frac{l^2}{m(a + x)^3} = -\frac{k}{(a + x)^n}. \tag{4.101}$$

Expanding both sides by the power series expansion and retaining first-order terms only, one arrives at

$$\ddot{x} - \frac{l^2}{m^2a^3}\left(1 - \frac{3x}{a}\right) = -\frac{k}{ma^n}\left(1 - \frac{nx}{a}\right). \tag{4.102}$$

Substitution from Eq. (4.97) gives

$$\ddot{x} + \frac{l^2}{m^2a^4}(3 - n)x = 0. \tag{4.103}$$

For Eq. (4.103) to be periodic, we once again get $n < 3$.

4.13 The Precessing Ellipse as Superposition of Two Power Laws

The examples of Sec. 4.8 indicate that when two types of forces act on the planet, the stronger force (the force with the higher power in r or lower inverse power in r) determines the basic nature of the orbit, with the weaker force (the force with the lower r-dependence) modifying it. In this section, we return to the forward problem of orbit determination under the inverse square law of gravitation when a perturbing force varying as the inverse cube of the distance is present. Here

$$f(r) = -\frac{GMm}{r^2} + \frac{a}{r^3}. \tag{4.104}$$

Inserting $f(r)$ in Eq. (4.79), we get

$$\frac{d^2u}{d\theta^2} + \left(1 + \frac{ma}{l^2}\right)u = \frac{GMm^2}{l^2}, \tag{4.105}$$

or

$$\frac{l^2}{l^2 + ma}\frac{d^2u}{d\theta^2} + u - \frac{GMm^2}{l^2 + ma} = 0. \tag{4.106}$$

Letting

$$x = u - \frac{GMm^2}{l^2 + ma}, \tag{4.107}$$

we have

$$\frac{d^2x}{d\theta^2} + \frac{l^2 + ma}{l^2}x = 0, \tag{4.108}$$

a simple harmonic motion with angular frequency

$$\omega = \sqrt{\frac{l^2 + ma}{l^2}}. \tag{4.109}$$

With a suitable choice of initial condition, we can take as solution to Eq. (4.108),

$$x = A \cos \omega \theta = A \cos \sqrt{\frac{l^2 + ma}{l^2}}. \tag{4.110}$$

Substituting back for x and u, we obtain the familiar equation of a conic (an ellipse, say)

$$r = \frac{p}{1 + e \cos \phi}, \tag{4.111}$$

where

$$\phi = \sqrt{\frac{l^2 + ma}{l^2}} \theta = \alpha \theta, \tag{4.112}$$

with semi-latus rectum

$$p = \frac{l^2 + ma}{GMm^2}, \tag{4.113}$$

and eccentricity

$$e = \frac{l^2 + ma}{GMm^2} A. \tag{4.114}$$

Equation (4.112) is the equation of an ellipse in the coordinate system $(r, \phi = \alpha \theta)$. Three special cases deserve attention. First, if $a = 0$, we get a *closed ellipse* of Keplerian motion given by Eqs. (4.17)–(4.19). Second, if $a < 0$, we have an attractive perturbing force. Then $\alpha < 1$, and θ leads ϕ. Thus, we have a *precessing ellipse*, whose line of apsides advances in time. Finally, when $a > 0$, we have a repulsive perturbing force. In that case, $\alpha > 1$, θ trails ϕ, and the apsidal line rotates in the retrograde sense (Fig. 4.4).

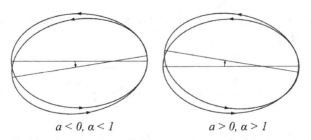

$$a < 0, \alpha < 1 \qquad\qquad a > 0, \alpha > 1$$

Fig. 4.4

4.14 The Precession of Mercury's Perihelion

It was known from the mid-nineteenth Century that the line of apsides of the planet Mercury advanced $574''$ of arc per century, out of which $531''$ was accounted for by Newtonian gravitational perturbations caused by the other planets, but the rest of $43''$ per century was unaccounted for. This phenomenon is well-known as the "perihelic precession of Mercury" (as the perihelion of a planet is observed more accurately than the aphelion). In 1915, Einstein provided the explanation of this exact amount on the basis of his General Theory of Relativity [cf. Eddington (1963)]. The perihelic precession of Mercury was considered the crown jewel of the General Theory of Relativity.

In Einstein's theory of gravitation, the inverse square force law of Newton is reproduced, as well as an additional term varying inversely as the fourth power of r:

$$f(r) = -\frac{GMm}{r^2} - \frac{a}{r^4}. \qquad (4.115)$$

The resulting differential equation of the orbit is [cf. Marion and Thornton (1995)]

$$\frac{d^2u}{d\theta^2} + u = \frac{GMm^2}{l^2} + \alpha u^2, \qquad (4.116)$$

where

$$\alpha = \frac{3GM}{c^2}, \qquad (4.117)$$

and c is the velocity of light in vacuum. Neglecting α, we get the Newtonian solution for the equation of a conic in the first approximation

$$u = \frac{1}{p}(1 + e \cos \theta), \qquad (4.118)$$

where p is the semi-latus rectum and e the eccentricity of the ellipse (we restrict our attention to the ellipse only). Substituting for u on the right-hand side of Eq. (4.116) and converting θ in the square term to 2θ, we have

$$\frac{d^2u}{d\theta^2} + u = \frac{1}{p} + \frac{\alpha}{p^2}\left[1 + 2e\cos\theta + \frac{e^2}{2}(1 + \cos 2\theta)\right]. \qquad (4.119)$$

Among all the additional terms in Eq. (4.119), only the term in $\cos \theta$ has the same periodicity as the first approximation solution and is to be retained. The rest of the terms can be discarded [cf. Eddington (1963)].

Now, the particular integral of the differential equation

$$\frac{d^2u}{d\theta^2} + u = \frac{2\alpha e}{p^2}\cos\theta, \qquad (4.120)$$

is

$$u_1 = \frac{\alpha e}{p}\theta \sin\theta. \qquad (4.121)$$

u_1 is now added to u in Eq. (4.118) to yield the second-order solution

$$u = \frac{1}{p}\left[1 + e\left(\cos\theta + \frac{\alpha}{p}\theta\sin\theta\right)\right]. \qquad (4.122)$$

Since α is small, by setting

$$d\theta = \frac{\alpha}{p}\theta, \qquad (4.123)$$

we have

$$u = \frac{1}{p} \left[1 + e \cos \left(\theta - d\theta \right) \right].$$ (4.124)

While the planet moves through one revolution (2π radians), the perihelion advances through an angle

$$d\theta = \frac{2\pi\alpha}{p}.$$ (4.125)

Putting the values of α and p from Eqs. (4.117) and (4.18), respectively, we get

$$d\theta = 6\pi \left(\frac{GMm}{cl} \right)^2.$$ (4.126)

Alternatively, from Eqs. (2.75) and (1.4), a more familiar form of the perihelic precession emerges:

$$d\theta = \frac{6\pi GM}{c^2 p} = \frac{6\pi GM}{c^2 a(1 - e^2)}.$$ (4.127)

Equations (4.127) indicates that the perihelic precession is greater for orbits having smaller semi-major axes and greater eccentricities. In the solar system, this is most noticeable for Mercury [cf. Marion and Thornton (1995)].

4.15 The Runge–Lenz Vector

Besides the conservation of the orbital angular momentum and the total energy, there exists another vector, commonly known as the **Runge–Lenz vector** (sic), which remains constant in planetary motion under the inverse square law of gravitation. It was Laplace, who was actually credited with the discovery of this vector [cf. Goldstein (1980)]. The Runge–Lenz vector is defined as

$$\vec{R} = \vec{p} \times \vec{l} - GMm^2\hat{r}, \tag{4.128}$$

where the notations have their usual meaning. To prove the constancy of $\vec{R}$ in time, it suffices to show that

$$\frac{d}{dt}(\vec{p} \times \vec{l}) = \frac{d}{dt}\left(GMm^2\hat{r}\right). \tag{4.129}$$

The left-hand side of Eq. (4.129) is

$$\frac{d}{dt}(\vec{p} \times \vec{l}) = \dot{\vec{p}} \times \vec{l} + \vec{p} \times \dot{\vec{l}} = \dot{\vec{p}} \times \vec{l}, \tag{4.130}$$

since $\vec{l}$ is a constant of motion. In Eq. (4.130), the dot on top of a symbol indicates total derivative with respect to time. By virtue of Newton's second law of motion and the law of gravitation, the rate of change of momentum is

$$\dot{\vec{p}} = -\frac{GMm}{r^3}\vec{r}. \tag{4.131}$$

Substituting for $\dot{\vec{p}}$ from Eq. (4.131) and the definition of $\vec{l}$ in Eq. (4.130), and expanding the resulting vector triple product, we get

$$\frac{d}{dt}(\vec{p} \times \vec{l}) = -\frac{GMm^2}{r^3}[\vec{r}(\vec{r} \cdot \dot{\vec{r}}) - \dot{\vec{r}}(\vec{r} \cdot \vec{r})]. \tag{4.132}$$

Now

$$\vec{r} \cdot \dot{\vec{r}} = \frac{1}{2}\frac{d}{dt}(\vec{r} \cdot \vec{r}) = \frac{1}{2}\frac{d}{dt}(r^2) = r\dot{r}. \tag{4.133}$$

Hence

$$\frac{d}{dt}(\vec{p} \times \vec{l}) = -\frac{GMm^2}{r^3}(r\dot{r}\vec{r} - r^2\dot{\vec{r}}). \tag{4.134}$$

We can expand the right-hand side of Eq. (4.129) as

$$\frac{d}{dt}\left(GMm^2\hat{r}\right) = \frac{d}{dt}\left(GMm^2\frac{1}{r}\vec{r}\right) = -\frac{GMm^2}{r^3}(r\dot{r}\vec{r} - r^2\dot{\vec{r}}). \tag{4.135}$$

Equations (4.134) and (4.135) prove the constancy of the Runge–Lenz vector $\vec{R}$.

To find the orientation of the Runge–Lenz vector, we first take its dot product with the invariant direction of the angular momentum [cf. Goldstein (1980)]:

$$\vec{R} \cdot \vec{l} = \left(\vec{p} \times \vec{l}\right) \cdot \vec{l} - \frac{GMm^2}{r}\vec{r} \cdot \vec{l} = 0. \qquad (4.136)$$

Thus, the Runge–Lenz vector is perpendicular to the angular momentum vector. In other words, it is in the orbital plane of the planet. Next, let ϕ denote the angle between the Runge–Lenz vector and the position vector. Then

$$\vec{R} \cdot \vec{r} = Ar\cos\phi = (\vec{p} \times \vec{l}) \cdot \vec{r} - \frac{GMm^2}{r}\vec{r} \cdot \vec{r}. \qquad (4.137)$$

Rearranging the first term on the right by cyclic permutation, we get

$$Rr\cos\phi = l^2 - GMm^2 r. \qquad (4.138)$$

Hence

$$r = \frac{\frac{l^2}{GMm^2}}{1 + \frac{R}{GMm^2}\cos\phi}. \qquad (4.139)$$

This is identical to the polar equation of the ellipse with ϕ replacing the angular coordinate θ and the eccentricity given by

$$e = \frac{R}{GMm^2}. \qquad (4.140)$$

Thus ϕ and θ are the same variable. Hence, the Runge–Lenz vector is directed along the major axis towards the perihelion, i.e., along the positive x-axis (Fig. 4.5). This actually constitutes an alternative derivation of the equation of orbit (Goldstein, 1980).

Equation (4.140) furnishes the magnitude of the Runge–Lenz vector in terms of the eccentricity of the orbit: $R = GMm^2e$. Some authors [e.g., Bate *et al.* (1971)] define the *eccentricity vector* as one parallel with the Runge–Lenz vector, having a magnitude equal to the eccentricity of the orbit,

$$\vec{e} = \frac{1}{GMm^2}\vec{R}. \tag{4.141}$$

One can define another vector,

$$\vec{S} = \vec{l} \times \vec{R}. \tag{4.142}$$

Then, $\vec{S}$ being perpendicular to two constant vectors, is itself a constant vector. It is directed along the latus rectum. One can identify this vector with the constant velocity vector $Ve\hat{y}$ discussed in Sec. 1.12. If we normalize the magnitudes of the three constant vectors $\vec{R}, \vec{S}$ and $\vec{l}$ to unity, then we have the three orthogonal unit vectors $(\hat{x}, \hat{y}, \hat{z})$ which form the basis of our rectangular coordinate system (Fig. 4.5).

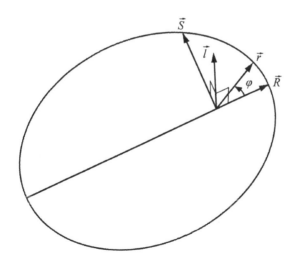

Fig. 4.5

Exercises

4.1. Show that the angular momentum of a particle is conserved in a general central force problem.

4.2. Show that the orbit of a particle subjected to a central force lies on a plane.

4.3. Deduce Kepler's three laws of planetary motion from Newton's law of gravitation.

4.4. Show that the planetary problem can be reduced to an equivalent one-dimensional problem in the radial coordinate.

4.5. Derive the orbit of a particle under the action of an attractive force varying inversely as the cube of the distance from the attracting center.

4.6. Derive Hamilton's equations of motion by equating the differentials of the Hamiltonian from Eqs. (4.40) and (4.42).

4.7. Derive Hamilton's equations of motion from Hamilton's principle (4.55).

4.8. Deduce the principle of least action from Hamilton's principle.

4.9. Deduce Jacobi's form of the principle of least action.

4.10. Verify Eq. (4.66).

4.11. Find the radial dependence of the central force under which a particle describes the hyperbolic spiral,

$$r = \frac{a}{\theta}.$$

4.12. Find the radial dependence of the central force under which a particle describes a trajectory given by the logarithmic spiral,

$$r = e^{a\theta}.$$

4.13. Find the radial dependence of the central force under which a particle describes a trajectory given by the Archimedean spiral,

$$r = a\theta.$$

4.14. Find the radial dependence of the central force under which a particle describes a trajectory given by the cardioid,

$$r = a(1 + \cos\theta).$$

4.15. Find the radial dependence of the central force under which a particle describes a trajectory given by the circle,

$$r = 2a\cos\theta.$$

4.16. Find the radial dependence of the central force under which a particle describes a trajectory given by the lemniscate of Bernoulli

$$r^2 = a^2 \cos 2\theta.$$

5

Vector Hodographs in Planetary Motion

5.1 The Hodograph

The *hodograph* is the locus of a vector drawn from a fixed point. It was first introduced by Hamilton in 1847. Hodographs have been used in dynamics [cf. Lamb (1961)], fluid mechanics (Courant and Friederichs, 1948), orbital mechanics (Altman, 1965) and population dynamics (Bartlett and Conklin, 1985). In particle dynamics, the hodograph of the position vector is nothing but the trajectory of the particle. More interesting aspects of the motion are revealed by the hodograph of the velocity vector.

In projectile motion under gravity, the hodograph of the position vector is, of course, a parabola. It is easy to see that the hodograph of the velocity vector is a vertical straight line. Since the acceleration vector in projectile motion is a constant vector, its hodograph is a point.

5.2 Vector Hodographs in Uniform Circular Motion

The uniform circular motion is a common type of motion which includes the motion of an electron around a positive charge, and also a limiting case of the Keplerian motion with eccentricity zero. Consider a uniform circular motion with a constant radius r. The speed v and angular speed ω are both constants and are related by

$$v = \omega r. \tag{5.1}$$

In plane polar coordinates (r, θ) in the orbital plane of motion with the attracting center as the center of coordinates, the position vector is written as

$$\vec{r} = r\hat{r}. \tag{5.2}$$

The rate of change of the unit vectors are, from Eqs. (1.21) and (1.22):

$$\frac{d\hat{r}}{dt} = \frac{d\theta}{dt}\hat{\theta} = \omega\hat{\theta}, \tag{5.3}$$

and

$$\frac{d\hat{\theta}}{dt} = -\frac{d\theta}{dt}\hat{r} = -\omega\hat{r}. \tag{5.4}$$

Differentiating Eq. (5.1) with respect to time and utilizing Eqs. (5.2)–(5.4), we get

$$\vec{v} = \frac{d\vec{r}}{dt} = \omega r\hat{\theta} = v\hat{\theta}, \tag{5.5}$$

$$\vec{a} = \frac{d\vec{v}}{dt} = -\omega v\hat{r} = -\omega^2 r\hat{r} = -\frac{v^2}{r}\hat{r}, \tag{5.6}$$

and

$$\vec{j} = \frac{d\vec{a}}{dt} = -\omega a\hat{\theta} = -\omega^3 r\hat{\theta} = -\frac{v^3}{r^2}\hat{\theta}. \tag{5.7}$$

In uniform circular motion, the position, velocity, acceleration and jerk vectors all have constant magnitudes and rotate at the same

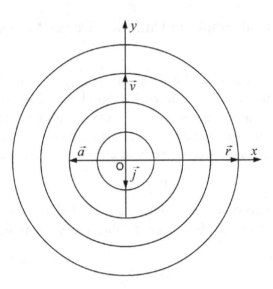

Fig. 5.1

constant angular velocity ω. Thus the hodographs of all four vectors are circles having different radii. However, the successive vectors lead their predecessor vectors by 90° and bear a ratio of ω in magnitude to them (Fig. 5.1).

It is possible to consider the hodographs of angular vectors. The angular displacement itself is a monotonically increasing quantity directed along the positive z-direction as given by the right-hand rule. As such, its hodograph is a straight line along the positive z-axis. The angular velocity vector is constant both in magnitude and direction:

$$\vec{\omega} = \omega\hat{z}. \tag{5.8}$$

Thus, its hodograph is a point on the positive z-axis at a distance of ω from the center. Since $\hat{z}$ is a constant vector, the angular acceleration vector is null vector:

$$\vec{\alpha} = \frac{d\vec{\omega}}{dt} = \vec{0}, \tag{5.9}$$

which is located at the center and has universal direction.

5.3 Orientation of the Orbital Ellipse

The conventional equation of an orbital ellipse having semi-latus rectum p in polar coordinates (r, θ) is written as

$$r = \frac{p}{1 + e\cos\theta}. \qquad (5.10)$$

Such an orbit has its perihelion point on the positive x-axis (a of Fig. 5.2). The orientation of this ellipse can be rotated counterclockwise by advancing the polar coordinate θ through any given angle. For example, by replacing θ by $\theta + \pi/2$, we get

$$r = \frac{p}{1 + e\sin\theta}. \qquad (5.11)$$

The perihelion point of this ellipse now lies on the positive y-axis (b of Fig. 5.2).

If we replace θ by $\theta + \pi$ in Eq. (5.10), we get

$$r = \frac{p}{1 - e\cos\theta}. \qquad (5.12)$$

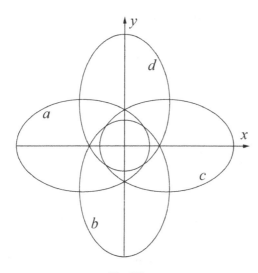

Fig. 5.2

The perihelion point of this ellipse lies on the negative x-axis (c of Fig. 5.2). Likewise, by replacing θ by $\theta + 3\pi/2$ in Eq. (5.10), one gets

$$r = \frac{p}{1 - e\sin\theta}. \tag{5.13}$$

The perihelion of the orbital ellipse now lies on the negative y-axis (d of Fig. 5.2). This goes to demonstrate that besides the conventional Eq. (5.10), other representations of the ellipse are possible. In the following sections, however, we restrict our discussion to the representation given by Eq. (5.10).

5.4 Polar Coordinates of Special Points on the Orbital Ellipse

The ellipse is a curve given by two parameters. Any suitable choice of two independent parameters will unambiguously define the ellipse. Commonly, (a and e) or (p and e) are chosen, even though other representations such as (a and b) are possible. We have frequently discussed the results of special points on the ellipse. They are the ends of the major axis (perihelion and aphelion points), ends of the minor axis and ends of the latera recta. The polar coordinates of these eight special points are summarized from the earlier chapters in Table 5.1.

5.5 Hodograph of the Position Vector in Planetary Motion

The hodograph of the position vector of a planet from the Sun is obviously the trajectory of the planet, which is an ellipse in accordance with Kepler's first law. The shape of the ellipse is entirely determined by the eccentricity, whereas the major axis is a measure of the size of the ellipse or the energy of the orbit. Figure 5.3 shows

Table 5.1. Polar coordinates of special points on the orbital ellipse.

Location of a point on the Ellipse	$r = r(p, e)$	$r = r(a, e)$	$\theta = \theta(e)$
Perihelion, end of major axis	$\frac{p}{1+e}$	$a(1 - e)$	0
End of latus rectum, moving away from Sun	p	$a(1 - e^2)$	$\pi/2$
End of minor axis, moving away from Sun	$\frac{p}{1-e^2}$	a	$\pi - \cos^{-1} e$
End of latus rectum through empty focus, moving away from Sun	$\frac{1+e^2}{1-e^2}p$	$a(1 + e^2)$	$\pi - \cos^{-1} \frac{2e}{1+e^2}$
Aphelion, end of major axis	$\frac{p}{1-e}$	$a(1 + e)$	π
End of latus rectum through empty focus, moving towards Sun	$\frac{1+e^2}{1-e^2}p$	$a(1 + e^2)$	$\pi + \cos^{-1} \frac{2e}{1+e^2}$
End of minor axis, moving towards Sun	$\frac{p}{1-e^2}$	a	$\pi + \cos^{-1} e$
End of latus rectum, moving towards Sun	p	$a(1 - e^2)$	$3\pi/2$

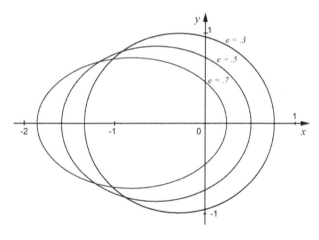

Fig. 5.3

three orbits having the same major axis (and therefore energy) but different eccentricities of $e = 0.3, 0.5$ and 0.7. According to Kepler's third law, the periods of the three orbits are also the same.

Figure 5.4 shows three elliptical orbits having the eccentricities $e = 0.3, 0.5$ and 0.7 but the same semi-latus rectum. In this case the

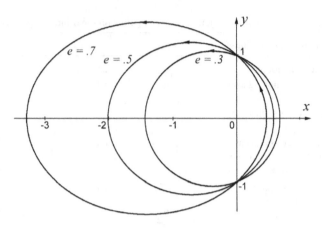

Fig. 5.4

energies of the orbits are different: they are proportional to $(1 - e^2)$. The higher the eccentricity, the greater is the energy.

5.6 Velocities at Special Points of an Elliptical Orbit

The velocity vector $\vec{v}$ is the first total derivative of the position vector $\vec{r}$. It is tangential to the orbit at any point. In polar coordinates, we have

$$\vec{v} = v_r \hat{r} + v_\theta \hat{\theta}. \qquad (5.14)$$

The radial and transverse components are, respectively, given by (from Sec. 1.12)

$$v_r = Ve \sin \theta, \qquad (5.15)$$

and

$$v_\theta = V(1 + e \cos \theta), \qquad (5.16)$$

where

$$V = \frac{l}{mp}. \qquad (5.17)$$

The magnitude of the velocity or speed is the scalar

$$v = \sqrt{v_r^2 + v_\theta^2}. \tag{5.18}$$

An elegant representation of the velocity components was given in Sec. 1.12, Eq. (1.78). In a mixed Cartesian and polar coordinate system, we wrote

$$\vec{v} = V\hat{\theta} + Ve\hat{y}, \tag{5.19}$$

where the first term represented the tangential velocity in a uniform circular motion having a magnitude V given by Eq. (5.17) and the second term was a constant velocity of magnitude Ve in the positive y-direction. Figure 5.5 shows the resolution of the velocity vector of the planet in accordance with Eq. (5.19) at special points on its orbit. Interestingly, the magnitudes of the two components of the velocity are the same at all points in the orbit.

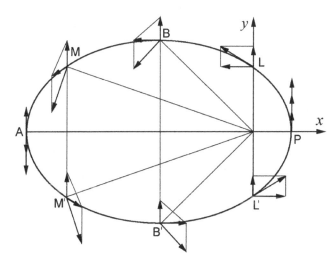

Fig. 5.5

5.7 Velocity Hodographs in Planetary Motion

In order to plot the velocity hodograph of the planet, we require the angle α the velocity vector makes with the positive x-axis. It can be verified that α leads θ according to the following equation (Tan, 1994):

$$\alpha = \theta + \tan^{-1} \frac{v_r}{v_\theta}. \tag{5.20}$$

Figure 5.6 shows the construction of the velocity hodograph for the Keplerian orbit of Fig. 5.5. Clearly, the hodograph is a circle of radius V, whose center is at a distance Ve from the orgin on the positive y-axis. A formal proof is found in Timoshenko and Young (1948).

The speeds of the planet can be evaluated easily from Fig. 5.6 and Eq. (5.19). By the straightforward application of the Pythagoras'

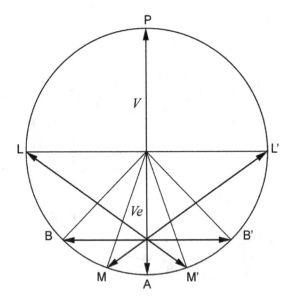

Fig. 5.6

theorem, we get

$$v_P = (1 + e)V, \tag{5.21}$$

$$v_A = (1 - e)V, \tag{5.22}$$

$$v_L = \sqrt{1 + e^2}\,V, \tag{5.23}$$

$$v_{L'} = \sqrt{1 + e^2}\,V, \tag{5.24}$$

$$v_B = \sqrt{1 - e^2}\,V, \tag{5.25}$$

and

$$v_{B'} = \sqrt{1 - e^2}\,V. \tag{5.26}$$

From the ends of the latus rectum through the empty focus, one has to employ the law of cosines and the expression for β from Eq. (1.12):

$$v_M = \frac{1 - e^2}{\sqrt{1 + e^2}}\,V, \tag{5.27}$$

and

$$v_{M'} = \frac{1 - e^2}{\sqrt{1 + e^2}}\,V. \tag{5.28}$$

Equations (5.21)–(5.28) are consistent with those obtained from Eq. (5.14). They also furnish the velocity theorems obtained in Chap. 1:

$$v_P v_A = v_B^2 = v_{B'}^2, \tag{5.29}$$

$$v_L v_{M'} = v_B^2 = v_{B'}^2, \tag{5.30}$$

and

$$v_M v_{L'} = v_B^2 = v_{B'}^2, \tag{5.31}$$

which are special cases of Eq. (1.56). One can verify that Eqs. (5.29)–(5.31) satisfy the sagittal theorem for the velocity hodograph circle of Fig. 5.6.

Figure 5.7 shows the velocity hodographs for the orbits of Fig. 5.4. They are all circles of radii V, but are displaced in the

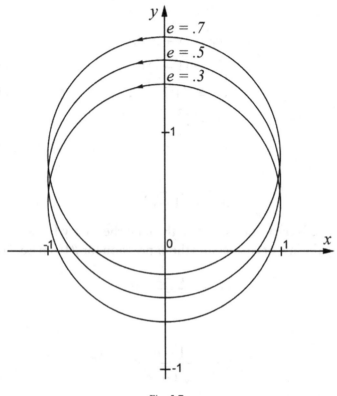

Fig. 5.7

positive y-direction by a distance equal to e times the radius V (Tan, 1994). This result follows from Eq. (5.19).

5.8 Acceleration Hodographs in Planetary Motion

Rewriting the expression for the velocity from Eqs. (5.14)–(5.17), we have

$$\vec{v} = \frac{le\sin\theta}{mp}\hat{r} + \frac{l(1 + e\cos\theta)}{mp}\hat{\theta}. \qquad (5.32)$$

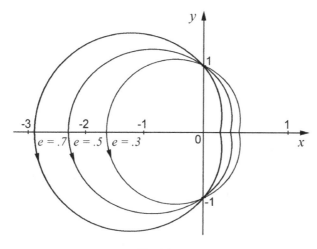

Fig. 5.8

Differentiating with respect to time and substituting for the derivatives of the unit vectors, we get

$$\vec{a} = -\frac{l^2}{m^2 p^3}(1 + e\cos\theta)^2 \hat{r}. \tag{5.33}$$

The acceleration vector, unlike the velocity vector, does not possess a transverse component and is anti-parallel to the position vector of the planet. It thus leads (or trails, for that matter) the position vector by π. Figure 5.8 depicts the hodographs of the acceleration vector of the orbits of Figs. 5.4 and 5.7 as obtained from Eq. (5.33). For higher eccentricities, the hodographs represent higher orders of Pascal's limacons (Altman, 1965).

5.9 Hodographs of the Jerk Vector in Planetary Motion

Differentiate Eq. (5.33) once more with respect to time and we get the jerk vector (Tan, 1994)

$$\vec{j} = j_r \hat{r} + j_\theta \hat{\theta}, \tag{5.34}$$

with

$$j_r = \frac{2l^3 e}{m^3 p^5} \sin\theta(1 + e\cos\theta)^3, \qquad (5.35)$$

and

$$j_\theta = -\frac{l^3}{m^3 p^5}(1 + e\cos\theta)^4. \qquad (5.36)$$

Since the jerk vector has a transverse component, one requires the angle β it makes with the positive x-axis in order to plot it. From Tan (1992b), we have

$$\beta = \theta - \frac{\pi}{2} - \tan^{-1}\frac{j_r}{j_\theta}. \qquad (5.37)$$

Figure 5.9 depicts the hodographs of the jerk vector for the orbits under study (Figs. 5.4, 5.7 and 5.8). The figure clearly indicates that the hodographs become more prominent with increasing eccentricity. It should be noted that whilst the curves appear to be elliptical, they are not ellipses.

5.10 Hodographs of Rotational Quantities in Planetary Motion

Vector hodographs may also be considered for rotational quantities in planetary motion, which are solenoidal vectors (Tan, 1994). Firstly, the areal velocity and the angular momentum of the planet are constant vectors whose directions are given by the right-hand rule. Their hodographs are, therefore, points on the positive z-axis.

Secondly, the angular velocity of the planet may be obtained from the expression for the angular momentum:

$$\vec{\omega} = \frac{l}{mp^2}(1 + e\cos\theta)\,\hat{z}, \qquad (5.38)$$

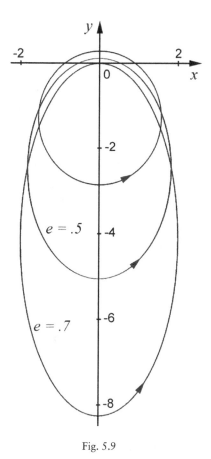

Fig. 5.9

where $\hat{z}$ is the unit vector in the positive z-direction, perpendicular to the plane of the orbit. Thirdly, the angular acceleration of the planet is obtained by differentiating Eq. (5.38) with respect to time [cf. Tan (1988)]:

$$\vec{\alpha} = -\frac{2l^2 e}{m^2 p^4} \sin \theta \, (1 + e \cos \theta)^3 \, \hat{z}. \qquad (5.39)$$

The hodographs of the angular velocity and the angular acceleration are less dramatic than those of their translational counterparts, being merely straight lines along the z-axis. The angular velocity hodographs lie entirely on the positive z-axis, whereas the angular acceleration hodographs extend on either side of the origin.

It is interesting to observe that the velocity, acceleration and jerk and angular velocity of the planet all attain their maximum and minimum values at the perihelion (where r is minimum) and aphelion (where r is maximum), respectively. In the limiting case of a circular orbit ($e = 0$), the hodographs of the position, velocity, acceleration and the jerk vectors are all circular and the phase difference between any two adjacent vectors is $\pi/2$. In this case, the angular velocity is constant and its hodograph is merely a point on the positive z-axis. The angular acceleration is then the null vector, whose position is at the origin.

Exercises

5.1. Obtain the hodograph of the position vector of a projectile under gravity.

5.2. Obtain the hodograph of the velocity vector of a projectile under gravity.

5.3. What is the hodograph of the acceleration vector of a projectile under gravity?

5.4. Prove that the hodograph of the velocity vector in planetary motion is a circle.

5.5. Verify Eq. (5.20).

5.6. Verify the speed theorems given by Eqs. (5.29)–(5.31) using the sagittal theorem in Fig. 5.6.

5.7. Derive Eq. (5.33).

5.8. Verify Eq. (5.37).

5.9. Derive Eq. (5.39).

Planetary Motion in Cartesian Coordinates

6

6.1 Cartesian Coordinates versus Polar Coordinates

In two-dimensional space, Cartesian coordinates and polar coordinates are the two most common and widely-used coordinate systems. Cartesian coordinates, like the rectangular coordinates in three-dimensional space, are stand-alone coordinate systems. The polar coordinates, on the other hand, must be defined relative to a Cartesian coordinates framework. In this sense, Cartesian coordinates are more fundamental than polar coordinates. Cartesian coordinates are used in general discussions, where no special symmetry is involved. Polar coordinates are widely-used in the discussion of plane polar curves [cf. Lockwood (1961) and Lawrence (1972)].

For the treatment of planetary motion, polar coordinates seem to be the exclusive choice of the coordinate system used, because of the natural advantages they bring over Cartesian coordinates.

Yet, Cartesian coordinates have been used in the past [e.g. Moulton (1970), Brouwer and Clemence (1961), Whittaker (1961), Timoschenko and Young (1948)]. In this chapter, we treat the planetary problem in Cartesian coordinates and explore the areas in which such coordinates provide an advantage over the polar coordinates.

6.2 Equations of an Ellipse in Cartesian Coordinates

The standard equation of the ellipse in polar coordinates with its right focus at the origin and its major axis in the x-direction is written as

$$r = \frac{p}{1 + e \cos \theta}.$$ (6.1)

The transformation equations from Cartesian to polar coordinates are given by

$$x = r \cos \theta,$$ (6.2)

and

$$y = r \sin \theta.$$ (6.3)

The inverse transformation relations are

$$r = \sqrt{x^2 + y^2},$$ (6.4)

and

$$\theta = \sin^{-1} \frac{y}{\sqrt{x^2 + y^2}} = \cos^{-1} \frac{x}{\sqrt{x^2 + y^2}} = \tan^{-1} \frac{y}{x}.$$ (6.5)

With the transformations (6.4) and (6.5), Eq. (6.1) assumes the form

$$\sqrt{x^2 + y^2} + ex = p.$$ (6.6)

Equation (6.6) is the standard form of the equation of an ellipse in Cartesian coordinates in terms of its eccentricity e and semi-latus

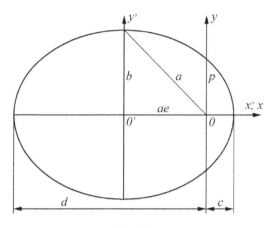

Fig. 6.1

rectum p. In the (a, e) representation, we have

$$\sqrt{x^2 + y^2} + ex = a(1 - e^2). \tag{6.7}$$

We can also proceed from the standard equation of the ellipse in Cartesian coordinates with its center at the origin (Fig. 6.1). In the (x', y') coordinates, we have

$$\frac{x'^2}{a^2} + \frac{y'^2}{b^2} = 1. \tag{6.8}$$

With the transformations

$$x' = x + ae, \tag{6.9}$$

and

$$y' = y, \tag{6.10}$$

we get

$$[(x + ae)^2 - a^2](1 - e^2) + y^2 = 0. \tag{6.11}$$

It can be verified that forms (6.7) and (6.11) are equivalent.

Table 6.1 summarizes the equations of the ellipse in both Cartesian and polar coordinates. Evidently, the equations are more

Table 6.1.   Equations of orbital ellipse.

Representation	Cartesian coordinates	Polar coordinates
(p, e)	$\sqrt{x^2 + y^2} + ex = p$	$r = \frac{p}{1 + e\cos\theta}$
(a, e)	$\sqrt{x^2 + y^2} + ex = a(1 - e^2)$	$r = \frac{a(1 - e^2)}{1 + e\cos\theta}$
(a, b)	$\dfrac{\left(x + \sqrt{a^2 - b^2}\right)^2}{a^2} + \dfrac{y^2}{b^2} = 1$	$r = \dfrac{b^2}{a + \sqrt{a^2 - b^2}\cos\theta}$
(c, d)	$(x - c)(x + d) + \dfrac{(c + d)^2}{4cd} y^2 = 0$	$r = \dfrac{2cd}{c + d + (d - c)\cos\theta}$

elegant in polar coordinates, in general, and the simplest in the (p, e) representation.

Next, we can express the equation of the ellipse in terms of a and b. One can verify that

$$ae = \sqrt{a^2 - b^2}. \tag{6.12}$$

Equations (6.8)–(6.10) then give

$$\frac{\left(x + \sqrt{a^2 - b^2}\right)^2}{a^2} + \frac{y^2}{b^2} = 1. \tag{6.13}$$

Finally, a representation of the ellipse is possible in terms of the perihelion and aphelion distances (Fig. 6.1):

$$c = a(1 - e), \tag{6.14}$$

and

$$d = a(1 + e). \tag{6.15}$$

Equation (6.11) transforms to (Tan, 1992a)

$$(x - c)(x - d) + \frac{(c + d)^2}{4cd} y^2 = 0. \tag{6.16}$$

6.3 Cartesian Coordinates of the Special Points on the Orbital Ellipse

It is far easier to write down the Cartesian coordinates of the special points of the orbital ellipse than the polar coordinates, especially at the ends of the minor axis and the latus rectum through the empty focus. Table 6.2 compiles the Cartesian coordinates of the above special points.

The Cartesian coordinates of the special points on the ellipse also assume fairly simple forms in terms of the perihelion and aphelion distances (Tan, 1992a), which are displayed in Table 6.3.

Table 6.2. Cartesian coordinates of special points on the orbital ellipse.

Location of a point on the Ellipse	$x = x(a, e)$	$y = y(a, e)$
Perihelion, end of major axis	$a(1 - e)$	0
End of latus rectum, moving away from Sun	0	$a(1 - e^2)$
End of minor axis, moving away from Sun	$-ae$	$a\sqrt{1 - e^2}$
End of latus rectum through empty focus, moving away from Sun	$-2ae$	$a(1 - e^2)$
Aphelion, end of major axis	$-a(1 + e)$	0
End of latus rectum through empty focus, moving towards Sun	$-2ae$	$-a(1 - e^2)$
End of minor axis, moving towards Sun	$-ae$	$-a\sqrt{1 - e^2}$
End of latus rectum, moving towards Sun	0	$-a(1 - e^2)$

Table 6.3. Cartesian coordinates of special points on the orbital ellipse.

Location of a point on the Ellipse	$x = x(c, d)$	$y = y(c, d)$
Perihelion, end of major axis	c	0
End of latus rectum, moving away from Sun	0	$\frac{2cd}{c+d}$
End of minor axis, moving away from Sun	$\frac{c-d}{2}$	$\sqrt{cd}$
End of latus rectum through empty focus, moving away from Sun	$c - d$	$\frac{2cd}{c+d}$
Aphelion, end of major axis	$-d$	0
End of latus rectum through empty focus, moving towards Sun	$c - d$	$-\frac{2cd}{c+d}$
End of minor axis, moving towards Sun	$\frac{c-d}{2}$	$-\sqrt{cd}$
End of latus rectum, moving towards Sun	0	$-\frac{2cd}{c+d}$

It is interesting to note that the semi-major axis a, the semi-minor axis b and the semi-latus rectum p of the ellipse represent the arithmetic mean (AM), the geometric mean (GM) and the harmonic mean (HM) of the apsidal distances c and d, respectively.

$$(AM) = a = \frac{c+d}{2}, \qquad (6.17)$$

$$(GM) = b = \sqrt{cd}, \qquad (6.18)$$

and

$$(HM) = p = \frac{2cd}{c+d}. \qquad (6.19)$$

It follows that (see Fig. 6.1)

$$ap = b^2 = cd. \qquad (6.20)$$

The geometric mean between c and d is also the geometric mean between the arithmetic and harmonic means!

6.4 Kepler's Law of Areas in Cartesian Coordinates

The elementary area swept by the planet at the Sun is given in polar coordinates as

$$dA = \frac{1}{2}r^2 d\theta. \qquad (6.21)$$

To transform the expression to Cartesian coordinates, we recall that

$$x = r\cos\theta, \qquad (6.2)$$

and

$$y = r\sin\theta. \qquad (6.3)$$

Taking differentials, we have

$$dx = dr\cos\theta - r\sin\theta\, d\theta, \qquad (6.22)$$

and

$$dy = dr \sin\theta + r \cos\theta \, d\theta. \tag{6.23}$$

Substituting from the above equations, we get

$$x dy - y dx = r^2 d\theta. \tag{6.24}$$

Thus, Eq. (6.21) in Cartesian coordinates assumes the form

$$dA = \frac{1}{2}(x dy - y dx), \tag{6.25}$$

whence, we have for the areal velocity

$$\frac{dA}{dt} = \frac{1}{2}\left(x\frac{dy}{dt} - y\frac{dx}{dt}\right) = \frac{1}{2}(x\dot{y} - y\dot{x}). \tag{6.26}$$

The law of areas demands that

$$x\dot{y} - y\dot{x} = const., \tag{6.27}$$

or, by differentiating with respect to time,

$$x\ddot{y} = y\ddot{x}. \tag{6.28}$$

Alternatively, the elementary area can be expressed as a cross-product:

$$d\vec{A} = \frac{1}{2}\left(\vec{r} \times d\vec{r}\right) = \frac{1}{2}\begin{vmatrix} \hat{x} & \hat{y} & \hat{z} \\ x & y & z \\ dx & dy & dz \end{vmatrix}. \tag{6.29}$$

In the $x - y$ plane, $z = 0$ and $dz = 0$. Thus

$$d\vec{A} = \frac{1}{2}\left(x dy - y dx\right)\hat{z}, \tag{6.30}$$

which is the vectorial version of Eq. (6.25).

To summarize, the constancy of the areal velocity in Cartesian coordinates is written as

$$\dot{A} = \frac{1}{2}\left(x\dot{y} - y\dot{x}\right) = const. \tag{6.31}$$

Likewise, the conservation of angular momentum in Cartesian coordinates assumes the form

$$m(x\dot{y} - y\dot{x}) = l = const. \tag{6.32}$$

6.5 Inverse-Square Law in Cartesian Coordinates

In Cartesian coordinates, the equations of motion in the x- and y-directions are, in accordance with Newton's law:

$$\ddot{x} = -\frac{GM}{r^2}\cos\theta = -\frac{GMx}{r^3}, \tag{6.33}$$

and

$$\ddot{y} = -\frac{GM}{r^2}\sin\theta = -\frac{GMy}{r^3}. \tag{6.34}$$

Equations (6.33) and (6.34) readily give

$$x\ddot{y} - y\ddot{x} = \frac{d}{dt}\left(x\dot{y} - y\dot{x}\right) = 0, \tag{6.35}$$

or

$$\dot{A} = \frac{1}{2}\left(x\dot{y} - y\dot{x}\right) = const. \tag{6.31}$$

and

$$m(x\dot{y} - y\dot{x}) = l = const. \tag{6.32}$$

Thus, Kepler's law of areas and the conservation of angular momentum are direct consequences of Newton's law of gravitation.

6.6 Velocity of a Planet in Cartesian Coordinates

The velocity of the planet in Cartesian coordinates can be obtained from the orbital equation and the conservation of angular momentum. We have

$$\sqrt{x^2 + y^2} + ex = p. \tag{6.6}$$

Differentiation with respect to time gives

$$x\dot{x} + y\dot{y} + e\dot{x}\sqrt{x^2 + y^2} = 0. \tag{6.36}$$

Eliminations between Eqs. (6.6), (6.32) and (6.36) yield the velocity components in Cartesian coordinates

$$v_x = \dot{x} = -\frac{l}{mp} \frac{y}{\sqrt{x^2 + y^2}}, \tag{6.37}$$

and

$$v_y = \dot{y} = \frac{l}{mp} \left(\frac{x}{\sqrt{x^2 + y^2}} + e \right), \tag{6.38}$$

whence

$$\vec{v} = -V\frac{y}{r}\hat{x} + V\frac{x}{r}\hat{y} + Ve\hat{y}, \tag{6.39}$$

with

$$V = \frac{l}{mp}. \tag{1.64}$$

Equation (6.39) gives us the velocity theorem of Sec. 1.12 in Cartesian coordinates.

Alternatively, we can proceed directly from the discussions of Sec. 1.12, where the velocity of the planet was expressed in polar coordinates as

$$\vec{v} = Ve(\sin\theta\hat{r} + \cos\theta\hat{\theta}) + V\hat{\theta}. \tag{1.77}$$

Recall that the relations between the unit vectors in Cartesian and polar coordinates are the following:

$$\hat{r} = \cos\theta\hat{x} + \sin\theta\hat{y}, \tag{1.23}$$

$$\hat{\theta} = -\sin\theta\hat{x} + \cos\theta\hat{y}, \tag{1.24}$$

$$\hat{x} = \cos\theta\hat{r} - \sin\theta\hat{\theta}, \tag{1.25}$$

and

$$\hat{y} = \sin\theta\hat{r} + \cos\theta\hat{\theta}. \tag{1.26}$$

With proper substitutions, Eq. (1.77) transforms into

$$\vec{v} = (-V\sin\theta\hat{x} + V\cos\theta\hat{y}) + Ve\hat{y}. \tag{6.40}$$

The first term on the right-hand side of Eq. (6.37) is recognized as a uniform circular motion with speed V in the counter-clockwise direction, whereas the second term represents a constant linear velocity of magnitude Ve in the positive y-direction. This is the stated theorem in Sec. 1.12. Applying Eqs. (6.2) and (6.3) to Eq. (6.40), we arrive at Eq. (6.39), which is Cartesian version of Eq. (1.77).

6.7 Velocities at Special Points of an Elliptical Orbit

In Cartesian coordinates, the orthogonal components of the velocity vector of the planet are given as

$$v_x = -V\frac{y}{\sqrt{x^2 + y^2}}, \tag{6.41}$$

and

$$v_y = V\frac{x}{\sqrt{x^2 + y^2}} + eV. \tag{6.42}$$

The magnitude of the velocity is the speed of the planet, which is, of course, the same in any representation:

$$v = \sqrt{v_x^2 + v_y^2}. \qquad (6.43)$$

The velocity components and the speed of the planet at the special points of the orbit in Table 5.1 are calculated in Cartesian coordinates using Eqs. (6.41)–(6.43) and gathered in Table 6.4. Comparisons with the results of Table 1.2 indicate slight advantages in Cartesian coordinates, especially at the ends of the minor axis and the latus rectum through the empty focus (Tan, 1992a). Figure 6.2 shows the compositions of the velocity components at the special points of the orbital ellipse. Clearly, the velocities at the ends of the major and minor axes have only single components, which represent advantages over compositions in polar coordinates.

The theorems on the speeds of the planet at special points can be verified from the entries of Table 6.4:

$$v_P v_A = v_B^2 = v_L v_M. \qquad (6.44)$$

Table 6.4.   Velocity components at special points on the ellipse.

Location of a point on the Ellipse	v_x	v_y	v
Perihelion, end of major axis	0	$V(1+e)$	$V(1+e)$
End of latus rectum, moving away from Sun	$-V$	Ve	$V\sqrt{1+e^2}$
End of minor axis, moving away from Sun	$-V\sqrt{1-e^2}$	0	$V\sqrt{1-e^2}$
End of latus rectum through empty focus, moving away from Sun	$-V\frac{1-e^2}{1+e^2}$	$-Ve\frac{1-e^2}{1+e^2}$	$V\frac{1-e^2}{\sqrt{1+e^2}}$
Aphelion, end of major axis	0	$-V(1-e)$	$V(1-e)$
End of latus rectum through empty focus, moving towards Sun	$V\frac{1-e^2}{1+e^2}$	$-Ve\frac{1-e^2}{1+e^2}$	$V\frac{1-e^2}{\sqrt{1+e^2}}$
End of minor axis, moving towards Sun	$V\sqrt{1-e^2}$	0	$V\sqrt{1-e^2}$
End of latus rectum, moving towards Sun	V	Ve	$V\sqrt{1+e^2}$

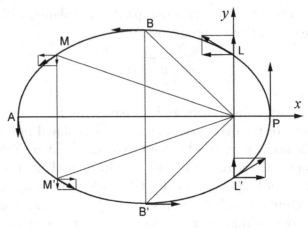

Fig. 6.2

6.8 The Radius of Curvature in Cartesian Coordinates

The expression for the radius of curvature in Cartesian coordinates is given by [cf. Gellert *et al.* (1977)]

$$\rho = \frac{\left[1 + \left(\frac{dy}{dx}\right)^2\right]^{3/2}}{\frac{d^2y}{dx^2}}. \tag{6.45}$$

For the orbitral ellipse, one can begin from the equation of the ellipse in Cartesian coordinates

$$\sqrt{x^2 + y^2} + ex = p. \tag{6.6}$$

Differentiate twice with respect to x to obtain the first and the second derivatives:

$$\frac{dy}{dx} = -\frac{x}{y} - \frac{e\sqrt{x^2 + y^2}}{y}, \tag{6.46}$$

and

$$\frac{d^2y}{dx^2} = \frac{e^2}{y} - \frac{1}{y}\left[1 + \left(\frac{dy}{dx}\right)^2\right]. \tag{6.47}$$

The radii of curvature at some of the special points can be found from Eqs. (6.45)–(6.47). At the end of the latus rectum L (Fig. 6.2), we have $x = 0$ and $y = p$. Here

$$\frac{dy}{dx} = -e, \tag{6.48}$$

$$\frac{d^2y}{dx^2} = -\frac{1}{p}, \tag{6.49}$$

and

$$\rho_L = -\left(1 + e^2\right)^{3/2} p. \tag{6.50}$$

The negative sign in Eq. (6.49) indicates concavity of the curve downwards at L. Equation (6.50) agrees with Eq. (2.10) but for the negative sign. At the other end of the latus rectum L' (Fig. 6.2), we have $x = 0$, but $y = -p$. In this case

$$\frac{dy}{dx} = e, \tag{6.51}$$

$$\frac{d^2y}{dx^2} = \frac{1}{p}, \tag{6.52}$$

and

$$\rho_{L'} = \left(1 + e^2\right)^{3/2} p. \tag{6.53}$$

The positive value of the second derivative signifies concavity upwards.

Similarly, at $M, x = -2ae, y = p$. Thus

$$\frac{dy}{dx} = e, \tag{6.54}$$

$$\frac{d^2y}{dx^2} = -\frac{1}{p}, \tag{6.55}$$

and

$$\rho_M = -\left(1 + e^2\right)^{3/2} p, \tag{6.56}$$

is in agreement with Eq. (6.50). But at $M', x = -2ae$, and $y = -p$. There

$$\frac{dy}{dx} = -e,\tag{6.57}$$

$$\frac{d^2y}{dx^2} = \frac{1}{p},\tag{6.58}$$

and

$$\rho_{M'} = \left(1 + e^2\right)^{3/2} p.\tag{6.59}$$

Two other points of interest are the ends of the minor axis. At B (Fig. 6.2), $x = -ae, y = b$, and $dy/dx = 0$. Following the procedure, we get

$$\rho_B = -\frac{p}{\left(1 - e^2\right)^{3/2}}.\tag{6.60}$$

At $B', x = -ae, y = -b$, and $dy/dx = 0$. Here

$$\rho_{B'} = \frac{p}{\left(1 - e^2\right)^{3/2}}.\tag{6.61}$$

The magnitudes of the radii of curvatures at the ends of the minor axis are consistent with Eq. (2.14):

$$\rho_B = \rho_{B'} = \frac{a^2}{b}.\tag{6.62}$$

It should be noted that at the ends of the major axis, dy/dx is infinite and the radii of curvatures become indeterminate. This is an area where Cartesian coordinates are at a great disadvantage compared with polar coordinates. In order to overcome this problem, it is necessary to interchange the abscissa and ordinate in Eq. (6.45), determine dx/dy, etc., and proceed from there (see Exercise 6.9).

The radius of curvature can alternatively be determined from its parametric form [cf. Gellert *et al.* (1977)]

$$\rho = \frac{\left[\left(\frac{dx}{du}\right)^2 + \left(\frac{dy}{du}\right)^2\right]^{3/2}}{\frac{dx}{du}\frac{d^2y}{du^2} - \frac{dy}{du}\frac{d^2x}{du^2}}. \qquad (6.63)$$

Equating the parameter u with time, we get a simplified version of the radius of curvature

$$\rho = \frac{v^3}{\dot{x}\ddot{y} - \dot{y}\ddot{x}}. \qquad (6.64)$$

The radii of curvatures at the special points (including the apsidal points) can be conveniently evaluated using Eq. (6.64).

The denominator of Eq. (6.64) can be simplified by direct substitutions from Eqs. (6.33), (6.34), (6.37) and (6.38), and using the equation of the ellipse (6.6), to:

$$\ddot{x}\ddot{y} - \dot{y}\ddot{x} = \frac{GMl}{mr^3}, \qquad (6.65)$$

whence from Eq. (2.3), the centripetal force on the planet

$$f_c = \frac{mv^2}{\rho} = \frac{GMl}{mr^3v}. \qquad (6.66)$$

The force of attraction of the Sun on the planet is given by Eq. (2.45):

$$f_g = \frac{mv^2}{\rho}\frac{v}{v_\theta} = \frac{GMl}{r^3v_\theta}. \qquad (6.67)$$

Recognizing that

$$rv_\theta = r^2\frac{d\theta}{dt} = \frac{l}{m}, \qquad (6.68)$$

one arrives at Newton's law of gravitation

$$f_g = \frac{GMm}{r^2}. \qquad (6.69)$$

We should caution that this is not a derivation of the inverse square law of gravitation, which was already assumed in Eqs. (6.33) and (6.34). Rather, it validates the chain of arguments used here.

6.9 The Equation of Energy in Cartesian Coordinates

The equation of energy can be obtained in Cartesian coordinates in a straightforward manner. Taking the self-dot product of the velocity from Eq. (6.39), we have

$$v^2 = \vec{v} \cdot \vec{v} = V^2 \left(\frac{y}{r}\right)^2 + V^2 \left(\frac{x}{r} + e\right)^2. \qquad (6.70)$$

On expanding and simplifying,

$$v^2 = 2V^2 \frac{p}{r} - V^2 \frac{p}{a}. \qquad (6.71)$$

By recalling that

$$V = \frac{l}{mp}, \qquad (1.64)$$

and

$$\frac{l^2}{m^2 p} = GM, \qquad (2.75)$$

we arrive at the equation of energy in Cartesian coordinates,

$$v^2 = GM \left(\frac{2}{\sqrt{x^2 + y^2}} - \frac{1}{a}\right). \qquad (6.72)$$

6.10 The Equation of an Orbit in Cartesian Coordinates

The conservation of angular momentum as given by Eq. (6.32) was a direct consequence of the inverse-square law of gravitation. Also, the velocity components as given by Eqs. (6.37) and (6.38) were obtained as applied to the elliptical orbit in either the rectangular or polar coordinate systems. When Eqs. (6.37) and (6.38) are substituted back into Eq. (6.32), the Cartesian equation of the orbital ellipse emerges.

$$\sqrt{x^2 + y^2} + ex = p. \tag{6.6}$$

One should note that this does not constitute a derivation of Kepler's first law in Cartesian coordinates. The formal derivation in Sec. 4.3 called for polar coordinates. A derivation of Kepler's first law in Cartesian coordinates appears in Timoshenko and Young (1948). However, that derivation uses substitutions which are tantamount to using polar coordinates.

6.11 Acceleration of a Planet in Cartesian Coordinates

The acceleration of the planet is obtained by the straightforward differentiation of Eq. (6.39):

$$\vec{a} = \frac{d\vec{v}}{dt} = -V\frac{x(x\dot{y} - y\dot{x})\hat{x}}{(x^2 + y^2)^{3/2}} - V\frac{y(x\dot{y} - y\dot{x})\hat{y}}{(x^2 + y^2)^{3/2}}. \tag{6.73}$$

Substituting for the angular momentum from Eq. (6.32) and V from Eq. (1.65), we get the acceleration of the planet in Cartesian coordinates

$$\vec{a} = -\frac{l^2}{m^2 p}\frac{x\hat{x} + y\hat{y}}{(x^2 + y^2)^{3/2}}, \tag{6.74}$$

which reaffirms that the force on the planet is attractive and is inversely proportional to its distance from the Sun. Substitution from Eq. (2.64) retrieves Newton's law of gravitation

$$\vec{f} = m\vec{a} = -\frac{GMm\hat{r}}{r^2}. \tag{2.38}$$

6.12 The Jerk Vector in Cartesian Coordinates

The jerk vector is found by differentiating equation (6.74) with respect to time:

$$\vec{j} = \frac{d\vec{a}}{dt} = j_x\hat{x} + j_y\hat{y}, \tag{6.75}$$

where

$$j_x = \frac{l^2}{m^2 p} \frac{(2x^2 - y^2)\dot{x} + 3xy\dot{y}}{(x^2 + y^2)^{5/2}}, \tag{6.76}$$

and

$$j_y = \frac{l^2}{m^2 p} \frac{(2y^2 - x^2)\dot{y} + 3xy\dot{x}}{(x^2 + y^2)^{5/2}}, \tag{6.77}$$

with $\dot{x}$ and $\dot{y}$ given by Eqs. (6.41) and (6.42), respectively. These expressions are more cumbersome than those given in polar coordinates and no advantage is gained here.

6.13 The Runge–Lenz Vector in Cartesian Coordinates

In Sec. 4.15, the existence of a conserved vector other than the angular momentum vector was discussed. The Runge–Lenz vector was defined as

$$\vec{R} = \vec{p} \times \vec{l} - GMm^2\hat{r}. \tag{4.128}$$

To prove the constancy of this vector, particularly its direction, was not a simple matter. One had to show that the total derivative of this vector with respect to time was zero. One had to further obtain a scalar equation in the form of the equation of the orbit in order to determine its direction. In rectangular coordinates (x, y, z), on the other hand, this becomes a straightforward exercise (Tan, 1992a).

We have from Eqs. (6.39) and (1.65),

$$\vec{p} = m\vec{v} = -\frac{l}{p}\frac{y}{r}\hat{x} + \frac{l}{p}\left(\frac{x}{r}+e\right)\hat{y}. \qquad (6.78)$$

Also,

$$\vec{l} = l\hat{z}. \qquad (6.79)$$

Thus

$$\vec{p} \times \vec{l} = \frac{l}{p}\begin{vmatrix} \hat{x} & \hat{y} & \hat{z} \\ -\frac{y}{r} & \frac{x}{r}+e & 0 \\ 0 & 0 & l \end{vmatrix} = \frac{l^2}{p}\left[\left(\frac{x}{r}+e\right)\hat{x}+\frac{y}{r}\hat{y}\right]. \qquad (6.80)$$

Furthermore, from Eq. (2.64), we have

$$GMm^2\hat{r} = GMm^2\frac{\vec{r}}{r} = \frac{l^2}{p}\left[\frac{x}{r}\hat{x}+\frac{y}{r}\hat{y}\right]. \qquad (6.81)$$

Hence, by Eqs. (6.80) and (6.81),

$$\vec{R} = \frac{l^2 e}{p}\hat{x} = GMm^2 e\hat{x}. \qquad (6.82)$$

Equation (6.82) immediately tells us that $\vec{R}$ is a constant vector directed along the positive x-direction.

6.14 Lagrange's Equations in Cartesian Coordinates

In the Lagrangian formulation of the planetary motion in Cartesian coordinates, the Lagrangian is written as

$$L = T - V = \frac{1}{2}m\dot{x}^2 + \frac{1}{2}m\dot{y}^2 + \frac{GMm}{(x^2+y^2)^{3/2}}. \qquad (6.83)$$

The Lagrange equations in the Cartesian coordinates x and y are

$$\frac{d}{dt}\left(\frac{\partial L}{\partial \dot{x}}\right) - \frac{\partial L}{\partial x} = 0, \qquad (6.84)$$

and

$$\frac{d}{dt}\left(\frac{\partial L}{\partial \dot{y}}\right) - \frac{\partial L}{\partial y} = 0. \tag{6.85}$$

Written explicitly, Eqs. (6.84) and (6.85) become

$$\ddot{x} + \frac{GMx}{(x^2 + y^2)^{3/2}} = 0, \tag{6.86}$$

and

$$\ddot{y} + \frac{GMy}{(x^2 + y^2)^{3/2}} = 0. \tag{6.87}$$

Equations (6.86) and (6.87) are coupled equations and are equivalent to Eqs. (6.33) and (6.34). Thereafter, the treatment of the planetary problem remains the same.

6.15 The Two-Body Problem

Regardless of their masses, any two bodies can always be put into stable Keplerian orbits around their center of mass. Figure 6.3 shows

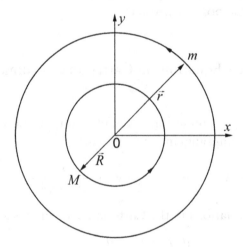

Fig. 6.3

two bodies of masses M and m moving in circular orbits around their center of mass, which is at the origin. If $\vec{R}$ and $\vec{r}$ represent the respective position vectors, then from the definition of the *center of mass*, we have

$$M\vec{R} + m\vec{r} = (M + m)\vec{0} = \vec{0}. \qquad (6.88)$$

Thus

$$\frac{\vec{R}}{\vec{r}} = -\frac{m}{M}. \qquad (6.89)$$

The two bodies must be at diametrically opposite locations at all times and the ratio of their distances from the center of mass is inversely proportional to their masses. Double differentiation of Eq. (6.88) with respect to time gives

$$M\ddot{\vec{R}} = -m\ddot{\vec{r}}, \qquad (6.90)$$

or,

$$\frac{\ddot{\vec{R}}}{\ddot{\vec{r}}} - \frac{m}{M}. \qquad (6.91)$$

From Eqs. (6.89) and (6.91), we arrive at the following theorem.

Theorem 6.1. *The accelerations of the bodies are directly proportional to their distances from the center of mass. The same result can be obtained from the expression of the centripetal acceleration $m\omega^2 r$, since ω is the same for both masses.*

If the two bodies are in elliptical orbits around their center of mass, the above equations still hold (Fig. 6.4). The larger body will move in a smaller ellipse. But the two ellipses will be similar and have the same eccentricity.

Consider the general two-body problem in space where two arbitrary masses m_1 and m_2 are in mutual orbits around each other. Let us assume that m_1 is greater than m_2, bearing in mind that this assumption poses no restriction on the generality of the problem. In a Cartesian coordinate system (x, y) in the plane of motion of the

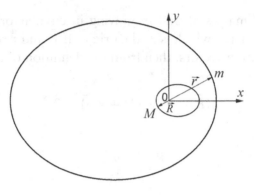

Fig. 6.4

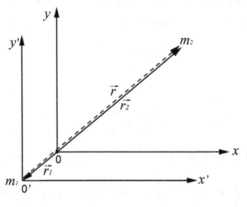

Fig. 6.5

two masses with the center of mass of the system at the origin O, the position vectors of m_1 and m_2 are $\vec{r}_1$ and $\vec{r}_2$, respectively (Fig. 6.5). Equation (6.88) then assumes the form

$$m_1\vec{r}_1 + m_2\vec{r}_2 = \vec{0}. \tag{6.92}$$

Furthermore, from vector addition, we have (Fig. 6.5),

$$\vec{r}_1 + \vec{r} = \vec{r}_2, \tag{6.93}$$

where $\vec{r}$ is the radius vector of m_2 from m_1. Solving for $\vec{r}_1$ and $\vec{r}_2$ in terms of $\vec{r}$ between Eqs. (6.92) and (6.93), we get

$$\vec{r}_1 = -\frac{m_2}{m_1 + m_2}\vec{r}, \tag{6.94}$$

and

$$\vec{r}_2 = \frac{m_1}{m_1 + m_2}\vec{r}. \tag{6.95}$$

Differentiation with respect to time yields the velocities of the two masses

$$\vec{v}_1 = \dot{\vec{r}}_1 = -\frac{m_2}{m_1 + m_2}\dot{\vec{r}}, \tag{6.96}$$

and

$$\vec{v}_2 = \dot{\vec{r}}_2 = \frac{m_1}{m_1 + m_2}\dot{\vec{r}}. \tag{6.97}$$

A second differentiation furnishes their accelerations

$$\vec{a}_1 = \ddot{\vec{r}}_1 = -\frac{m_2}{m_1 + m_2}\ddot{\vec{r}}, \tag{6.98}$$

and

$$\vec{a}_2 = \ddot{\vec{r}}_2 = \frac{m_1}{m_1 + m_2}\ddot{\vec{r}}. \tag{6.99}$$

By Newton's law of gravitation, the equation of motion of mass m_2 is given by

$$m_2\ddot{\vec{r}}_2 = -\frac{Gm_1m_2}{r^2}\hat{r}. \tag{6.100}$$

By Newton's third law of motion, the equation of motion of mass m_1, is likewise, given by

$$m_1\ddot{\vec{r}}_1 = \frac{Gm_1m_2}{r^2}\hat{r}. \tag{6.101}$$

Substitution of either Eq. (6.99) into Eq. (6.100) or Eq. (6.98) this Eq. (6.101) gives

$$\mu\ddot{\vec{r}} = -\frac{G(m_1 + m_2)\mu}{r^2}\hat{r}, \tag{6.102}$$

where

$$\mu = \frac{m_1 m_2}{m_1 + m_2} \qquad (6.103)$$

is the **reduced mass** of the system.

Equation (6.102) has the same form as Eq. (6.100) in a relative coordinate system (x_1, y_1) in which the mass m_1 is replaced by $(m_1 + m_2)$ at the origin of the coordinates (Fig. 6.5), and m_2 is replaced by the reduced mass of the system. Thus, a two-body problem is reduced to an equivalent one-body problem, when the radial distance between the two masses is used, instead of from the center of mass. This result is generally valid as long as Newton's third law of motion is obeyed [cf. Norwood (1979)].

The Lagrangian of the system is given by

$$L = T - V = \frac{1}{2} m_1 \dot{\vec{r}}_1 \cdot \dot{\vec{r}}_1 + \frac{1}{2} m_2 \dot{\vec{r}}_2 \cdot \dot{\vec{r}}_2 - \frac{G m_1 m_2}{r}. \qquad (6.104)$$

Substituting from Eqs. (6.96), (6.97) and (6.103), we obtain

$$L = \frac{1}{2} \mu \dot{\vec{r}} \cdot \dot{\vec{r}} - \frac{G(m_1 + m_2)\mu}{r}. \qquad (6.105)$$

Lagrange's equations give the same equation of motion (6.102) and the solution is the same as that of a one-body problem, with the mass of the larger body replaced by the total mass of the system and the mass of the revolving body replaced by the reduced mass of the system.

If the two masses are in circular orbit around the center of mass as in Fig. 6.3, we get, by equating the centripetal force on m_2 with the gravitational force:

$$\frac{m_2 v_2^2}{r_2} = \frac{G m_1 m_2}{r^2}. \qquad (6.106)$$

Substituting from Eq. (6.95), we get

$$v_2 = \sqrt{\frac{G m_1^2}{r(m_1 + m_2)}}. \qquad (6.107)$$

The period of m_2 about the center of mass is obtained by substituting once more from Eq. (6.95) [cf. Lim (1995)]:

$$P_2 = \frac{2\pi r^{3/2}}{\sqrt{G(m_1 + m_2)}}.$$ (6.108)

The angular velocity of m_2 is

$$\omega_2 = \frac{2\pi}{P_2} = \frac{\sqrt{G(m_1 + m_2)}}{r^{3/2}}.$$ (6.109)

The period and angular velocity of m_1 about the center of mass are obviously the same as those of m_1. Squaring Eq. (6.108), one obtains Kepler's third law for the two-body problem.

6.16 The Three-Body Problem

Since the time of Kepler and Newton, the three-body problem had engaged the minds of the greatest mathematicians like Euler, Lagrange, Jacobi, Poincare, and others. It can be safely stated that the solution to the general three-body problem has not been found and probably does not exist. However, several special cases of the three-body problem are known to exist. In 1765, Euler obtained periodic solutions for three colinear masses, the ratios of whose distances were fixed. But *Euler's colinear solutions of the three-body problem* are unstable and are therefore are not expected to be found in nature. In 1772, Lagrange found stable solutions for three masses which were located at the vertices of an equilateral triangle. Many Trojan asteroids observed leading and trailing Jupiter in its orbit around the Sun constitute exmples of *Lagrange's equilateral triangle solution of the three-body problem.*

Recently, an interesting new solution of the three-body problem was discovered by Moore (1993) who found that three equal masses can move in a figure eight curve. Exact solutions to Moore's discovery were obtained by Chenciner and Montgomery (2000).

Computer simulations by Simo have demonstrated that the **eight-figure solution of the three-body problem** is probably a stable one. Moreover, computations reveal that there exists a finite probability that such a system could be found somewhere in the universe, and perhaps even in our own galaxy.

An elegant treatment of the equilateral triangle solution of the three-body problem is found in Finlay-Freundlich (1958). Consider three masses m_1, m_2 and m_3 moving around a common center of mass in a plane. Let the Cartesian coordinates of the masses be (x_1, y_1), (x_2, y_2) and (x_3, y_3) respectively (Fig. 6.6). If the masses are at the corners of an equilateral triangle, then the coordinates of the center C of the triangle are given by

$$x_C = \frac{x_1 + x_2 + x_3}{3}, \tag{6.110}$$

and

$$y_C = \frac{y_1 + y_2 + y_3}{3}. \tag{6.111}$$

The coordinates of the center of mass CM are

$$x_{CM} = \frac{m_1 x_1 + m_2 x_2 + m_3 x_3}{m_1 + m_2 + m_3}, \tag{6.112}$$

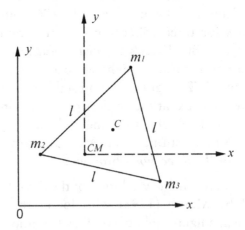

Fig. 6.6

and

$$y_{CM} = \frac{m_1 y_1 + m_2 y_2 + m_3 y_3}{m_1 + m_2 + m_3}. \tag{6.113}$$

If h is the height of the triangle and l the length of its side, then the distance of a mass (m_1, say) from the center of the triangle is given by

$$d_1 = \frac{2}{3}h = \frac{l}{\sqrt{3}}, \tag{6.114}$$

where

$$d_1 = (x_C - x_1)^2 + (y_C - y_1)^2. \tag{6.115}$$

Similarly, the distance of the same mass from the center of mass is given by

$$D_1 = (x_{CM} - x_1)^2 + (y_{CM} - y_1)^2. \tag{6.116}$$

It can be shown from Eqs. (6.110)–(6.116) that [cf. Finlay-Freundlich (1958)]

$$D_1 = \frac{\sqrt{m_2^2 + m_2 m_3 + m_3^2}}{m_1 + m_2 + m_3} l. \tag{6.117}$$

If the center of mass is reckoned as the origin of coordinates, then

$$m_1 x_1 + m_2 x_2 + m_3 x_3 = 0, \tag{6.118}$$

and

$$m_1 y_1 + m_2 y_2 + m_3 y_3 = 0. \tag{6.119}$$

Consider the motion of mass m_1. The x-component of force on m_1 due to m_2 and m_3 is, according to Eq. (6.33):

$$f_{1x} = -Gm_1 m_2 \frac{x_1 - x_2}{r_{12}^2} - Gm_1 m_3 \frac{x_1 - x_3}{r_{13}^2}. \tag{6.120}$$

Since $r_{12} = r_{13} = r_{23} = l$, we have from Eq. (6.117),

$$f_{1x} = -GM_1 m_1 \frac{x_1}{r_1^3}, \tag{6.121}$$

where

$$M_1 = \frac{\left(m_2^2 + m_2 m_3 + m_3^2\right)^{3/2}}{(m_1 + m_2 + m_3)^2} \tag{6.122}$$

is the effective mass with which m_2 and m_3 attract m_1 towards the center of mass of the system [cf. Finlay-Freundlich (1958)]. Similar results hold for the y-component of the force on m_1 and the x- and y-components of forces of m_2 and m_3. This demonstrates the equilateral triangular solution of the three-body problem of Lagrange.

An alternative approach to this problem is found in Lim (1994). Without loss of generality, we can choose the coordinate system such that the three masses m_1, m_2 and m_3 are located at $(0, 0)$, $(l, 0)$ and $(l/2, \sqrt{3}l/2)$, respectively (Fig. 6.7). The position vector of the center of mass is then

$$\vec{r}_{CM} = \frac{\left(m_2 + \frac{m_3}{2}\right)\hat{x} + \frac{\sqrt{3}}{2}\hat{y}}{m_1 + m_2 + m_3} l. \tag{6.123}$$

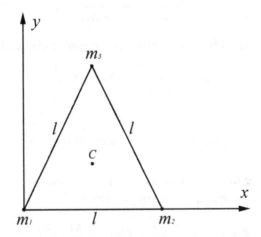

Fig. 6.7

The resultant gravitational force on m_1 due to m_2 and m_3 is

$$\vec{f}_1 = -\frac{Gm_1m_2}{l^2}\hat{x} - \frac{Gm_1m_3}{l^2}\left(\frac{\hat{x}}{2} + \frac{\sqrt{3}\hat{y}}{2}\right). \tag{6.124}$$

The magnitude of the force is

$$f_1 = \frac{Gm_1}{l^2}\sqrt{m_2^2 + m_2m_3 + m_3^2}. \tag{6.125}$$

Also, from Eq. (6.123), we arrive at the same expression as (6.116):

$$r_{CM} = D_1 = \frac{\sqrt{m_2^2 + m_2m_3 + m_3^2}}{m_1 + m_2 + m_3}. \tag{6.126}$$

Substituting from Eq. (6.126), Eq. (6.125) assumes the form

$$f_1 = -\frac{GM_1m_1}{r_{CM}^2}, \tag{6.127}$$

which is equivalent to Eq. (6.121).

If m_1 is in a circular orbit around the center of mass, its velocity v_1 can be obtained by equating (6.125) to the centripetal force [cf. Lim (1994)]

$$\frac{m_1v_1^2}{r_{CM}} = \frac{Gm_1}{l^2}\sqrt{m_2^2 + m_2m_3 + m_3^2}, \tag{6.128}$$

giving

$$v_1^2 = \frac{G}{l}\frac{m_2^2 + m_2m_3 + m_3^2}{m_1 + m_2 + m_3}. \tag{6.129}$$

The period of revolution of m_1 around the center of mass is then

$$P = \frac{2\pi r_{CM}}{v_1} = \frac{2\pi l^{3/2}}{\sqrt{G(m_1 + m_2 + m_3)}}, \tag{6.130}$$

which is also equal to those of m_2 and m_3. Note the similarity with Eq. (6.109).

The rectilinear solution of the three-body problem of Euler can be obtained from the theorem of Sec. 6.15. If the three bodies lie on a straight line and if the center of mass of the system is at the origin, then

$$m_1\vec{r}_1 + m_2\vec{r}_2 + m_3\vec{r}_3 = \vec{0}. \qquad (6.131)$$

This satisfies the conditions of dynamical equilibrium [Eqs. (6.118) and (6.119)]. Given the locations of any two masses, there are three possible locations of the third: between the two masses and outside the two on the same straight line [cf. Finlay-Freundlich (1958)]. However, the rectilinear solution is unstable because any radial motion of the masses will cause them to move on a path of no return.

Exercises

6.1. Verify the equivalence of Eqs. (6.7) and (6.11).

6.2. Derive Eq. (6.16).

6.3. Verify the entries of Table 6.2.

6.4. Verify the entries of Table 6.3.

6.5. Derive Eq. (6.25).

6.6. Derive Eqs. (6.37) and (6.38).

6.7. Verify the entries of Table 6.4.

6.8. Derive Eq. (6.40).

6.9. Calculate the radii of curvature at the perihelion and the aphelion by interchanging x and y in Eq. (6.45) and differentiating Eq. (6.6) twice with respect to y.

6.10. Calculate the radii of curvature at the perihelion and the aphelion using Eq. (6.64).

6.11. Derive Eq. (6.72).

6.12. Derive Eq. (6.74).

6.13. Derive Eq. (6.75).

6.14. Derive Eqs. (6.86) and (6.87).

6.15. Show that the path of a particle under an elastic force is an ellipse with its center at the origin.
6.16. Verify Eq. (6.102).
6.17. Prove the relation (6.117).
6.18. Derive Eq. (6.122).

7

Planetary Problem in Complex Coordinates

7.1 Complex Numbers

A *complex number* is a general number containing both a real and an imaginary part as represented by

$$z = a + ib, \qquad (7.1)$$

where a and b are both real and $i = \sqrt{-1}$. The real and imaginary parts of z are the following:

$$\text{Re}\,\{z\} = a, \qquad (7.2)$$

and

$$\text{Im}\,\{z\} = b. \qquad (7.3)$$

A real number is a special case of a complex number whose imaginary part is zero ($b = 0$). A purely imaginary number is a complex number whose real part is zero ($a = 0$), but whose complex part is not zero ($b \neq 0$). Thus, the conventional zero ($a = 0, b = 0$) is considered a real number.

The **modulus** or absolute value of a complex number is its magnitude as defined by

$$|z| = |a + ib| = \sqrt{a^2 + b^2}. \qquad (7.4)$$

The **complex conjugate** or simply conjugate of a complex number z is defined by

$$z^* = a - ib. \qquad (7.5)$$

z^* has the same modulus as z:

$$|z^*| = |a - ib| = \sqrt{a^2 + b^2}. \qquad (7.6)$$

Also, for any complex number z, z^*z and zz^* are real and equal:

$$z^*z = zz^* = a^2 + b^2. \qquad (7.7)$$

7.2 Graphical Representation of Complex Numbers

Just as all real numbers are represented as points on an infinite straight line, complex numbers are represented by points on an infinite plane. Thus z is represented by the point (a, b) in the x–y plane (Fig. 7.1). Likewise, z^* is represented by the point $(a, -b)$ in

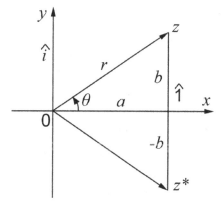

Fig. 7.1

the same plane. z and z^* are thus mirror images of one another about the real axis (Fig. 7.1). The graphical representation of complex numbers is called the **Argand diagram.**

7.3 Polar Coordinate Representation of Complex Numbers

The polar coordinates (r, θ) provide another convenient represent-ation of complex numbers. From Fig. 7.1,

$$z = x + iy = r\cos\theta + ir\sin\theta = re^{i\theta}, \qquad (7.8)$$

where we have used the **Euler formula**

$$e^{i\theta} = \cos\theta + i\sin\theta. \qquad (7.9)$$

Similarly, we have

$$z^* = x - iy = r\cos\theta - ir\sin\theta = re^{-i\theta}. \qquad (7.10)$$

As expected, Eqs. (7.8) and (7.10) give

$$z^*z = zz^* = r^2. \qquad (7.11)$$

7.4 Complex Numbers as Coplanar Vectors

There are many similarities between complex numbers and vectors in a two-dimensional plane. The situations are parallel if one equates the unit vectors in the x- and y-directions with 1 and i, respectively [cf. Morse and Feshbach (1953)]. The conditions of equality, addition and subtraction of complex numbers become identical to those of coplanar vectors [cf. Spiegel (1964)].

Figure 7.2 demonstrates the graphical addition and subtraction of complex numbers deemed as two-dimensional vectors. The addition of two vectors z_1 and z_2 obeys the **parallelogram law of**

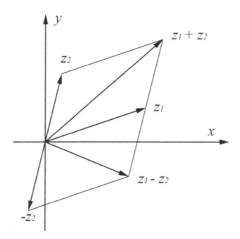

Fig. 7.2

forces. Evidently, this addition is **commutative**, i.e.,

$$z_1 + z_2 = z_2 + z_1. \qquad (7.12)$$

The subtraction of the vector z_2 from z_1 is performed by adding the negative of the former to the latter (Fig. 7.2):

$$z_1 - z_2 = z_1 + (-z_2). \qquad (7.13)$$

The equivalence of the product rules between complex numbers and coplanar vectors can also be demonstrated as follows: consider two complex numbers

$$z_1 = x_1 + iy_1, \qquad (7.14)$$

and

$$z_2 = x_2 + iy_2. \qquad (7.15)$$

Then

$$z_1^* = x_1 - iy_1, \qquad (7.16)$$

and

$$z_1^* z_2 = (x_1 x_2 + y_1 y_2) + i(x_1 y_2 - x_2 y_1). \qquad (7.17)$$

The real part of Eq. (7.17) constitutes the **scalar product** or **dot product** (also known as the **inner product**) in two-dimensional space, whereas the imaginary part of Eq. (7.17) comprises the **vector product** or **cross product** in two-dimensional space [cf. Spiegel (1964)].

$$z_1 \cdot z_2 = \text{Re}\{z_1^* z_2\} = x_1 x_2 + y_1 y_2, \qquad (7.18)$$

and

$$z_1 \times z_2 = \text{Im}\{z_1^* z_2\} = x_1 y_2 - x_2 y_1. \qquad (7.19)$$

Equation (7.17) is recognized as the **outer product** or **total product** (also called the **Kronecker product**) in two-dimensional space

$$z_1 \otimes z_2 = z_1^* z_2 = z_1 \cdot z_2 + i z_1 \times z_2. \qquad (7.20)$$

7.5 Polar Coordinates Representation of Coplanar Vectors

In plane polar coordinates, a radial vector is represented by

$$\vec{f} = f e^{i\theta}, \qquad (7.21)$$

where f is the magnitude of that vector, or modulus of the representative complex number. The complex conjugate of the radial vector is given by

$$\vec{f}^* = f e^{-i\theta}. \qquad (7.22)$$

Likewise, a tangential vector is represented by

$$\vec{f} = i f e^{i\theta}, \qquad (7.23)$$

and its complex conjugate is given by

$$\vec{f}^* = -i f e^{-i\theta}. \qquad (7.24)$$

As examples, the unit vectors in the radial and tangential directions are given by

$$\hat{r} = e^{i\theta}, \qquad (7.25)$$

and

$$\hat{\theta} = ie^{i\theta}. \tag{7.26}$$

Along any direction, we obtain Eqs. (1.21) and (1.22) directly:

$$\frac{d\hat{r}}{d\theta} = \hat{\theta}, \tag{1.21}$$

and

$$\frac{d\hat{\theta}}{d\theta} = -\hat{r}. \tag{1.22}$$

Furthermore, we obtain Eqs. (5.3) and (5.4) easily:

$$\frac{d\hat{r}}{dt} = \frac{d\theta}{dt}\hat{\theta}, \tag{5.3}$$

and

$$\frac{d\hat{\theta}}{dt} = -\frac{d\theta}{dt}\hat{r}. \tag{5.4}$$

7.6 Rotation of a Vector in Complex Coordinates

If a vector $\vec{f}$ is rotated through an infinitesimal angle $d\theta$, the change of the vector produced by the rotation in the forward direction is given by [cf. Morse and Feshbach (1953)]

$$d\vec{f} = if d\theta. \tag{7.27}$$

By separating the variables and integrating, one obtains the new vector

$$\vec{f} = fe^{i\theta}. \tag{7.21}$$

Thus, the operator $e^{i\theta}$ rotates a vector through a finite angle of θ in the forward (counter-clockwise) direction (Morse and

Feshbach, 1953). Similarly, the operator $e^{-i\theta}$ would rotate a vector in the backward (clockwise) direction. The vector in Eq. (7.23) is the result of rotation of the vector in Eq. (7.21) by an angle of $\pi/2$.

7.7 Central Force Motion in Complex Coordinates

Many dynamical problems take place in a two-dimensional plane which includes projectile motion under gravity, simple pendulum motion and central force motions. The central force problem has been elegantly treated in complex coordinates (Boyd and Raychowdhury, 1985; Finkel, 1990).

Consider the motion of a particle of mass m acted upon by a central force directed towards the origin and given in complex coordinates by

$$\vec{f} = fe^{i\theta}. \tag{7.21}$$

The position vector of the particle is

$$\vec{r} = re^{i\theta}. \tag{7.28}$$

By successive differentiations with respect to time, we get the velocity and the acceleration vectors

$$\vec{v} = \dot{r}e^{i\theta} + ir\dot{\theta}e^{i\theta}, \tag{7.29}$$

and

$$\vec{a} = (\ddot{r} - r\dot{\theta}^2)e^{i\theta} + i(2\dot{r}\dot{\theta} + r\ddot{\theta})e^{i\theta}. \tag{7.30}$$

Applying Newton's law, we have from Eqs. (7.27) and (7.30):

$$m(\ddot{r} - r\dot{\theta}^2)e^{i\theta} + im(2\dot{r}\dot{\theta} + r\ddot{\theta})e^{i\theta} = fe^{i\theta}. \tag{7.31}$$

Multiplying by $e^{-i\theta}$ and equating the real and imaginary parts, we obtain the equations of the radial and transverse components of motion [cf. Boyd and Raychowdhury (1985)]:

$$m(\ddot{r} - r\dot{\theta}^2) = f, \tag{7.32}$$

and

$$m(2\dot{r}\dot{\theta} + r\ddot{\theta}) = 0. \tag{7.33}$$

The advantage of this treatment is that it is not necessary to resort to Eqs. (5.3)–(5.4) above [cf. Finkel (1990)].

7.8 Conservation of Energy in Complex Coordinates

The total energy of a particle in the central force problem is

$$E = \frac{1}{2}m\vec{v} \cdot \vec{v} + V(r) = \frac{1}{2m}\vec{p} \cdot \vec{p} + V(r), \tag{7.34}$$

where the potential energy $V(r)$ is given by the equation

$$f = -\frac{dV(r)}{dr}. \tag{7.35}$$

Differentiating Eq. (7.34) with respect to time and applying Newton's law,

$$\frac{dE}{dt} = \frac{1}{2m}\left(\dot{\vec{p}} \cdot \vec{p} + \vec{p} \cdot \dot{\vec{p}}\right) + \frac{dV(r)}{dr}\dot{r} = \frac{1}{m}\vec{p} \cdot \vec{f} - f(r)\dot{r}. \tag{7.36}$$

Now, from Eq. (7.29),

$$\vec{p} = m\vec{v} = m(\dot{r}e^{i\theta} + ir\dot{\theta}e^{i\theta}). \tag{7.37}$$

Thus

$$\vec{p}^* = m\vec{v}^* = m(\dot{r}e^{-i\theta} - ir\dot{\theta}e^{-i\theta}), \tag{7.38}$$

and

$$\vec{p} \cdot \vec{f} = \text{Re}\{\vec{p}^*\vec{f}\} = m\dot{r}f. \tag{7.39}$$

Hence Eq. (7.36) gives

$$\frac{dE}{dt} = 0, \tag{7.40}$$

which states the conservation of total energy [cf. Finkel (1990)].

7.9 Conservation of Angular Momentum in Complex Coordinates

In complex coordinates, the angular momentum can be expressed as

$$\vec{l} = \vec{r} \times \vec{p} = \text{Im}\{\vec{r}^*\vec{p}\}. \qquad (7.41)$$

Therefore, the rate of change of angular momentum is [cf. Finkel (1990)]

$$\frac{d\vec{l}}{dt} = \frac{d}{dt}\text{Im}\{\vec{r}^*\vec{p}\} = \text{Im}\left\{\frac{d}{dt}(\vec{r}^*\vec{p})\right\} = \text{Im}\{m\vec{v}^*\vec{v} + \vec{r}^*\vec{f}\}. \qquad (7.42)$$

It can be shown that the argument of the right-hand side of Eq. (7.42) is real. In polar form, we have, from Eq. (7.29),

$$\vec{v}^*\vec{v} = (\dot{r}e^{-i\theta} - ir\dot{\theta}e^{-i\theta})(\dot{r}e^{i\theta} + ire^{i\theta}) = \dot{r}^2 + r^2\dot{\theta}^2. \qquad (7.43)$$

Furthermore, from Eqs. (7.22) and (7.28),

$$\vec{r}^*\vec{f} = r^{-i\theta}f^{i\theta} = rf. \qquad (7.44)$$

Thus, Eq. (7.42) demonstrates the conservation of angular momentum in central force motion.

7.10 Conservation of the Runge–Lenz Vector in Complex Coordinates

In Sec. 4.15, the Runge–Lenz vector was introduced as

$$\vec{R} = \vec{p} \times \vec{l} - GMm^2\hat{r}, \qquad (7.45)$$

where the symbols have their usual meaning. By inspection, $\vec{p} \times \vec{l}$ is nothing but the vector $l\vec{p}$ rotated by $-\pi/2$ and is thus $-il\vec{p}$ (Finkel,

1990). Thus, we have

$$\frac{d\vec{R}}{dt} = -il\frac{d\vec{p}}{dt} - GMm^2\frac{d\hat{r}}{dt}. \tag{7.46}$$

Now, from Eqs. (7.28) and (7.29),

$$\vec{l} = \text{Im}\{\vec{r}^*\vec{p}\} = \text{Im}\{mr\dot{r} + imr^2\dot{\theta}\} = mr^2\dot{\theta}. \tag{7.47}$$

Also, by Newton's law of gravitation,

$$\frac{d\vec{p}}{dt} = \vec{f} = -\frac{GMm}{r^2}\hat{r}. \tag{7.48}$$

Inserting Eqs. (7.47) and (7.48) in Eq. (7.46) and applying Eqs. (7.25) and (7.26), we arrive at

$$\frac{d\vec{R}}{dt} = iGMm^2\dot{\theta}\hat{r} - GMm^2\dot{\theta}\hat{\theta} = 0. \tag{7.49}$$

Equation (7.49) illustrates the conservation of the Runge–Lenz vector in planetary motion. The traditional treatment calls for the expansion of a vector triple product (Finkel, 1990).

Exercises

7.1. For two complex numbers $z_1 = a_1 + b_1i$ and $z_2 = a_2 + b_2i$, prove the following:

(a) $z_1 = z_2$ if $a_1 = a_2$ and $b_1 = b_2$,
(b) $z_1 + z_2 = (a_1 + a_2) + (b_1 + b_2)i$,
(c) $z_1 - z_2 = (a_1 - a_2) + (b_1 - b_2)i$,
(d) $z_1z_2 = (a_1a_2 - b_1b_2) + (a_1b_2 + a_2b_1)i$,
(e) $\frac{z_1}{z_2} = \frac{a_1a_2+b_1b_2}{a_2{}^2+b_2{}^2} - \frac{a_1b_2-a_2b_1}{a_2{}^2+b_2{}^2}i$, and
(f) $z_1z_2 = z_2z_1$ (commutative law of multiplication).

7.2. For complex numbers $z_1 = r_1e^{i\theta_1}$, $z_2 = r_2e^{i\theta_2}, \ldots, z_n = r_ne^{i\theta_n}$, prove the following:

(a) $z_1z_2 = r_1r_2e^{i(\theta_1+\theta_2)}$,
(b) $\frac{z_1}{z_2} = \frac{r_1}{r_2}e^{i(\theta_1-\theta_2)}$, and

(c) $z_1 z_2 \cdots z_n = r_1 r_2 \cdots r_n e^{i(\theta_1 + \theta_2 + \cdots + \theta_n)}$.

7.3. For a complex number $z = re^{i\theta} = r(\cos\theta + i\sin\theta)$, prove the following:

   (a) $z^n = r^n e^{in\theta} = r^n(\cos n\theta + i\sin n\theta)$ (**DeMoivre's Theorem**),

   (b) $\frac{1}{z} = \frac{1}{r}e^{-i\theta} = \frac{1}{r}(\cos\theta - i\sin\theta)$, and

   (c) $\frac{1}{z^n} = \frac{1}{r^n}e^{-in\theta} = \frac{1}{r^n}(\cos n\theta - i\sin n\theta)$.

7.4. Show that

   (a) $z^{\frac{1}{n}} = r^{\frac{1}{n}}e^{i\frac{\theta+2k\pi}{n}} = r^{\frac{1}{n}}\left(\cos\frac{\theta+2k\pi}{n} + i\sin\frac{\theta+2k\pi}{n}\right)$, where $k = 0, 1, 2, \ldots, n-1$,

   (b) $z^{\frac{1}{2}} = r^{\frac{1}{2}}e^{i\frac{\theta+2k\pi}{2}}$, where $k = 0, 1$, and

   (c) $1^{\frac{1}{n}} = e^{\frac{2k\pi i}{n}} = \cos\frac{2k\pi}{n} + i\sin\frac{2k\pi}{n}$, where $k = 0, 1, 2, \ldots, n-1$.

   Note that the n roots of unity are distributed uniformly on a unit circle with the center at the origin on an Argand diagram. The equation of the unit circle is $|z| = 1$.

7.5. Verify Eq. (7.20).

7.6. Verify Eqs. (5.3) and (5.4).

7.7. Verify Eqs. (7.32) and (7.33).

7.8. Verify Eq. (7.40).

7.9. Verify the conservation of angular momentum.

7.10. Verify Eq. (7.49).

Keplerian Motion in
the Solar System

8

8.1 Keplerian Motion in the Solar System

In the solar system, Keplerian motion is observed at two levels. Firstly, the planets, asteroids and comets revolve around the Sun in Keplerian orbits with the Sun at one of the foci. Secondly, satellites revolve around many planets, once again, obeying the laws of Kepler. No natural objects are known to revolve around satellites, as of yet, which means that Keplerian motion in the third level does not exist naturally in the solar system. However, man-made spacecraft can always be put into Keplerian orbits around the satellites.

In Sec. 2.2, two conditions of the applicability of Keplerian motion were set forth: (i) The distance between the two bodies was infinitely greater than their dimensions. (ii) The mass of the larger body was infinitely greater than that of the smaller body. In the following sections, we shall consider the applicability of Keplerian motion when one of the above conditions is not met.

8.2 Attraction of a Thin Uniform Spherical Shell on a Point Mass

Consider a thin uniform spherical shell of mass M, radius a, thickness t and density ρ and a particle of mass m at a distance r from the center of the shell (Fig. 8.1). We have

$$M = 4\pi a^2 t\rho. \tag{8.1}$$

Next, consider a circular segment of the shell of width $ad\theta$, each part of which subtends a cone of half-angle θ at the center of the shell and another cone of half-angle α at m. Then the mass of the circular segment is

$$dM = (2\pi a \sin\theta)(ad\theta)(t)(\rho) = 2\pi\rho ta^2 \sin\theta\, d\theta. \tag{8.2}$$

By symmetry, the gravitational force of dM on m is directed towards the center of the shell and has the magnitude (vide Fig. 8.1):

$$dF = \frac{GdMm}{x^2}\cos\alpha = \frac{GdMm}{x^2}\frac{r - a\cos\theta}{x}. \tag{8.3}$$

In Eq. (8.3), the two variables θ and x are connected by the law of cosines:

$$x^2 = r^2 + a^2 - 2ra\cos\theta. \tag{8.4}$$

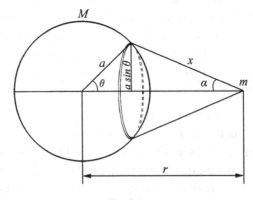

Fig. 8.1

By differentiation, we have

$$\sin\theta \, d\theta = \frac{x dx}{ra}. \tag{8.5}$$

Substituting Eqs. (8.2), (8.4) and (8.5) in Eq. (8.3), we get

$$dF = \frac{Gm\pi\rho t a}{r^2} \frac{r^2 - a^2 + x^2}{x^2} dx. \tag{8.6}$$

Integrating throughout the shell gives the net force of the shell M on m:

$$F = \frac{Gm\pi\rho t a}{r^2} \int_{r-a}^{r+a} \left(\frac{r^2 - a^2}{x^2} + 1 \right) dx = \frac{GMm}{r^2}. \tag{8.7}$$

Thus, we have the following theorem.

Theorem 8.1. *The gravitational force of a uniform thin spherical shell on a point mass is the same as if the mass of the entire shell were concentrated at its center.*

8.3 Attraction of a Spherically-Symmetric Body on a Point Mass

To the first approximation, the largest objects in the solar system are spherical in shape and possess spherical symmetry. These objects include the Sun, the planets, the large satellites, the largest asteroids and quite likely the largest cometary objects.

A spherically symmetric body can be regarded as the aggregate of a large number of uniform thin spherical shells having radii ranging from zero up to the radius of the sphere. Let the masses of these spherical shells be $m_1, m_2, m_3, \ldots, m_n$. If M is the mass of the spherical body, then

$$M = m_1 + m_2 + m_3 + \cdots + m_n = \sum_{i=1}^{n} m_i. \qquad (8.8)$$

Now, let there be a point mass m located at a distance r from the center of the sphere. Then, according to the above theorem, the attractive force of a general ith shell on the point mass m is directed towards the center and has the magnitude

$$F_i = \frac{Gm_i m}{r^2}, \quad i = 1, 2, 3, \ldots, n. \qquad (8.9)$$

By summing up over all the shells, the net attractive force of the spherical body of mass M on the point mass m is

$$F = \frac{Gm}{r^2}(m_1 + m_2 + m_3 + \cdots + m_n) = \frac{Gm \sum_{i=1}^{n} m_i}{r^2} = \frac{GMm}{r^2}. \qquad (8.10)$$

Thus, we have a corollary to the above theorem.

Corollary 8.1. *The gravitational force of a spherically symmetrical body on a point mass is the same as if the entire mass of the sphere were concentrated at its center.*

8.4 Physical Properties of the Earth and the Moon

The physical properties of the Earth and the Moon relevant to orbital motion are given in Tables 8.1 and 8.2, respectively. A heavenly body possessing appreciable mass and rotation (e.g. the Earth) assumes the shape of an oblate spheroid. The **volumetric radius** of such a body refers to the radius of a sphere having the same volume as the oblate spheroid. In the absence of sufficient rotation, the heavenly body (e.g. the Moon) assumes a more or less spherical shape.

For angular motion, the **sidereal period** refers to an inertial frame of reference (see Sec. 10.2). From our perspective, the distant

Table 8.1. Earth data.

Quantity	Value	Reference
Mass M	5.977414×10^{24} kg	Nelson and Loft (1962)
Gravitational parameter GM	3.986016×10^{14} m^3/s^2	Nelson and Loft (1962)
Equatorial radius R_e	6,378 km	McBride and Gilmour (2003)
Polar radius R_p	6,357 km	McBride and Gilmour (2003)
Volumetric radius $\sqrt[3]{R_e^2 R_p}$	6,371 km	McBride and Gilmour (2003)
Mean density	5.51×10^3 kg/m^3	McBride and Gilmour (2003)
Semi major axis a	149.6×10^6 km	McBride and Gilmour (2003)
Sidereal orbital period	365.256 days	McBride and Gilmour (2003)
Sidereal rotation period P_s	23.9345 h	New Solar System (1999)
Solar day	24 h	
Eccentricity e	0.017	McBride and Gilmour (2003)
Obliquity/axial inclination	23.45°	New Solar System (1999)
Ellipticity $\frac{R_e - R_p}{R_p}$	0.0034	New Solar System (1999)
Equatorial surface gravity g_e	9.78 m/s^2	New Solar System (1999)
Equatorial escape velocity V_e	11.2 km/s	New Solar System (1999)

Table 8.2. Moon data.

Quantity	Value	Reference
Mass	7.349×10^{22} kg	New Solar System (1999)
Equatorial radius	1,738 km	McBride and Gilmour (2003)
Polar radius	1,738 km	McBride and Gilmour (2003)
Mean radius	1,738 km	McBride and Gilmour (2003)
Mean density	3.34×10^3 kg/m^3	McBride and Gilmour (2003)
Sidereal period	27.322 days	New Solar System (1999)
Synodic period	29.53 days	Danby (1988)
Mean distance from the earth	384,400 km	New Solar System (1999)
Eccentricity	0.055	McBride and Gilmour (2003)
Obliquity/Axial inclination	6.67°	New Solar System (1999)
Ellipticity	0.002	New Solar System (1999)
Orbital inclination from ecliptic	5.2°	McBride and Gilmour (2003)
Equatorial surface gravity	1.62 m/s^2	New Solar System (1999)
Equatorial escape velocity	2.4 km/s	New Solar System (1999)

stars can be assumed to adequately represent an inertial frame. The sidereal period is considered to be the true period. For example, the sidereal rotation period of the Earth is 23 h 56 min 4 s or 23.9345 h. Due to the Earth's revolution, the Sun appears to rise and set once in 24 h at the equator, which defines a *solar day*.

The *synodic period* of the Moon is the time in which the Moon appears in the same position in the sky relative to the Sun, as observed from the Earth. Thus, it is the period after which the Moon exhibits the same phase to the Earth.

8.5 Velocities of Artificial Earth Satellites

The velocities of artificial Earth satellites can be conveniently obtained from elementary considerations. By virtue of the corollary of Sec. 8.3, the orbit of a satellite can be assumed to be Keplerian, with the center of the Earth at one focus. For simplicity's sake, we shall consider circular orbits here. For an orbit of radius R, we equate the centripetal force with the gravitational force:

$$-\frac{mv^2}{R} = -\frac{GMm}{R^2}.$$

(8.11)

Then

$$v = \sqrt{\frac{GM}{R}}.$$

(8.12)

The velocity of the satellite in a circular orbit depends only on the mass of the Earth M and radius of the orbit R. Theoretically, the smallest possible orbit is one, which barely skims the surface of the Earth. In practice, however, air resistance will rapidly consume any orbit below about 140 km above the surface. Nonetheless, we begin with the theoretically smallest orbit by setting R equal to the reference radius of the Earth. The values of M and R used are shown in Table 8.1. In Orbital Mechanics, the equatorial radius of the Earth is customarily taken as the reference radius [cf. Nelson and Loft (1962)]. The velocity of a satellite 'skimming' the surface of the Earth is then

$$v_0 = \sqrt{\frac{GM}{R_e}}.$$

(8.13)

Plugging in the values of G, M and R_e from Table 8.1, we get $v_0 \approx$ 7.91 km/s or 28,500 kph (17,700 mph). This represents the smallest velocity that the launch vehicle must attain before it can put a satellite in orbit.

On the surface of the Earth, locations at the equator have the maximum tangential velocity v_1. Referred to the quantities in Table 8.1,

$$v_1 = \frac{2\pi R_e}{P_s}.$$
(8.14)

v_1 has a magnitude of 1,670 kph (1,040 mph) and is directed eastward. This velocity can be deducted free of charge from v_0, if the satellite is launched eastward from an equatorial site. This is precisely the advantage gained by the French launch site at Kourou (latitude 5.1°) in French Guiana.

8.6 Periods of Artificial Earth Satellites

The period of a satellite in a circular orbit of radius R can be determined from its velocity. From Eq. (8.12), we get

$$P = \frac{2\pi R}{v} = \frac{2\pi R^{3/2}}{\sqrt{GM}}.$$
(8.15)

For example, for a typical *low Earth orbit* (LEO) satellite at an altitude of 300 km, the period is 90.5 min. Notice the form of Kepler's third law in Eq. (8.15), discussed earlier in Sec. 2.4.

A satellite having a revolution period exactly equal to the sidereal rotation period of the Earth (23.9345 h) is called a *geosynchronous satellite*. Such a satellite appears to oscillate about a fixed point on the equator. A geosynchronous satellite in the equatorial plane of the Earth is called a *geostationary satellite*. A geostationary satellite hovers over the same point on the equator and is ideal for satellite communication and continuous monitoring of the Earth's surface. From Eq. (8.15), we can find out the radial distance of a geostationary satellite from the center of the Earth:

$$R = \sqrt[3]{\frac{P^2 GM}{4\pi^2}}. \tag{8.16}$$

Letting $P = P_s$ in Eq. (8.16), one obtains $R \approx 42,164\,\text{km}$ or 6.61085 R_e. Thus, a geostationary satellite is situated at 6.61085 equatorial Earth radii, or $35,786$ km above the equator.

8.7 Velocity of Escape from the Surface of the Earth

The velocity of escape of a spacecraft from the Earth's surface can also be found from elementary considerations. It is the parabolic velocity discussed in Sec. 2.9. For a satellite in elliptical orbit around the Earth, the nearest approach of the satellite to the Earth is called the *perigee* and the farthest retreat of the satellite from the Earth is called the *apogee*. For a spacecraft to escape the gravitational field of the Earth, i.e., to never return, it is sufficient to assume that the kinetic energy of the spacecraft at an infinite distance from the Earth is zero. Furthermore, the potential energy of the spacecraft is also zero at infinity. From the law of conservation of energy, we can equate the total energies at a perigee point near the surface of the Earth and at infinity.

$$\frac{1}{2}mv_e^2 - \frac{GMm}{R_e} = 0 + 0. \tag{8.17}$$

We thus obtain the velocity of escape

$$v_e = \sqrt{\frac{2GM}{R_e}}. \tag{8.18}$$

From Eq. (8.13), we get

$$v_e = \sqrt{2}v_0. \tag{8.19}$$

The escape velocity from the surface of the Earth is $v_e \approx 11.18$ km/s, which is approximately 40,260 kph or 25,000 mph.

The escape velocity depends only upon the mass and radius of the attracting body. The escape velocity from the surface of the Moon is considerably less and is about 2.4 km/s. The escape velocity from the surface of the Sun, on the other hand, is about 1.4 million kilometers per hour. Such a speed represents the parabolic velocity near the surface of the Sun with which an unpowered spacecraft must travel in order to escape the solar system unaided by the gravity boosts from any planet.

8.8 Time of Travel to the Moon

The time of space travel from the Earth to the Moon can, once again, be determined from elementary considerations of Keplerian motion. The geometry of rendezvous of the spacecraft and the Moon is shown in Fig. 8.2. For the spacecraft, a highly elliptical orbit is the most economical. The powered section of the spaceflight is a small segment near the Earth. Hence a degenerate ellipse having the Earth and the Moon at two foci can be used. For the Moon's trajectory (eccentricity 0.055), a cicular orbit suffices. Then, the semi-major axis of the spacecraft orbit is half of that of the Moon's orbit. If we denote the periods of spacecraft's orbit and the Moon's orbit by P_S and P_M respectively, by Kepler's third law,

$$\frac{P_S^2}{P_M^2} = \frac{\left(\frac{a}{2}\right)^3}{a^3} = \frac{1}{8}. \tag{8.20}$$

The required time of flight (one way trip to the Moon) is

$$\frac{P_S}{2} = \frac{P_M}{4\sqrt{2}}. \tag{8.21}$$

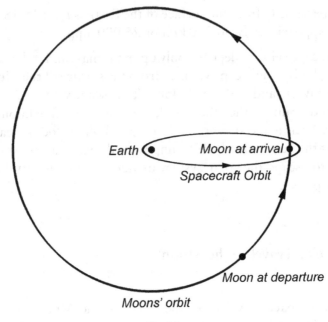

Fig. 8.2

Equating P_M to the sidereal revolution period of the Moon (27.3 days), we get 4.83 days for the time of flight to the Moon aboard a Keplerian trajectory.

8.9 Times of Travel to Mars and Venus

The easiest and most economical way for a spacecraft to transfer from one circular orbit to another coplanar circular orbit is via a cotangential elliptical orbit whose perigee/perihelion lies on the inner orbit and apogee/aphelion lies on the outer orbit (Fig. 8.3). This was first recognized by Hohmann and the orbit is called a **Hohmann transfer orbit**. Artificial Earth satellites in high altitude orbits (e.g. geosynchronous satellites) are normally placed there using Hohmann transfer orbits [cf. Bate *et al.* (1971)]. The satellite is first placed on a low altitude 'parking orbit'. A 'velocity boost' of

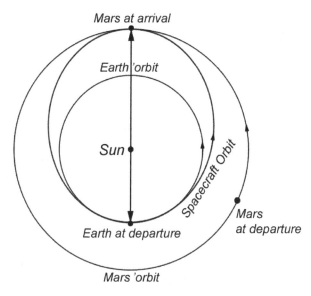

Fig. 8.3

the proper amount in the forward direction puts it in the transfer orbit. After it reaches the apogee point of the elliptical orbit, another velocity boost in the forward direction puts it the intended larger circular orbit [cf. Bate *et al.* (1971)].

In the solar system, all the planets (with the exception of Pluto) are in nearly coplanar orbits, which lie within 7° of the ecliptic. For a spacecraft journey from the Earth to Mars, the Hohmann transfer orbit is used, since it is the most economical of all transfer orbits [cf. Bate *et al.* (1971)]. The time of flight can be deduced from the properties of the transfer orbit, whose period is [from Eq. (2.70)]

$$P = \frac{2\pi}{\sqrt{GM}} a^{\frac{3}{2}}. \tag{8.22}$$

The time of flight (one way journey) is exactly half of this period. Also, the semi-major axis of the ransfer orbit is the arithmetic mean of those of the Earth and Mars:

$$a = \frac{a_E + a_M}{2}. \tag{8.23}$$

From Eqs. (8.22) and (8.23), the time of flight of the spacecraft from the Earth to Mars is

$$T = \frac{P}{2} = \frac{\pi}{\sqrt{GM}} \frac{(a_E + a_M)^{3/2}}{2\sqrt{2}}. \tag{8.24}$$

Putting the values of $a_E = 149.6 \times 10^6$ km and $a_M = 227.9 \times 10^6$ km, we get $T \approx 259$ days [cf. Marion and Thornton (1995)]. It should be noted that the Hohmann transfer orbit is also the slowest transfer orbits between two coplanar circular orbits [cf. Bate *et al.* (1971)]. For manned spaceflights to Mars in the future, faster orbits will be called for.

Hohmann orbits can also be used for transfer from a larger circular orbit to a smaller coplanar circular orbit [cf. Bate *et al.* (1971)]. In this case, the orbital velocity corrections are applied in the retrograde direction for the spacecreaft to plunge towards the Sun. Using the same analysis in reverse, we obtain the time of flight of a spacecraft from the Earth to Venus (semi-major axis 108.2×10^6 km) as 146 days [cf. Bate *et al.* (1971)]. Likewise, the time of flight between the Earth and Mercury (semi-major axis 57.91×10^6 km) is 105.5 days [Bate *et al.* (1971)]. It is faster to fall from the Earth to Mercury than from the Earth to Venus. But this is not surprising because Mercury is actually closer to the Earth than Venus at *opposition* unlike at *conjunction*. The transfer orbit to Mercury is entirely inside that of Venus.

8.10 The Mass of the Sun from the Period of the Earth

The mass of the Sun can be accurately determined from the sidereal period of a planet when the conditions for a one-body problem are met. From Eq. (2.61), we get

$$M = \frac{4\pi^2}{GP^2} a^3. \tag{8.25}$$

Since the mass of the Sun is over 330,000 times that of the Earth and the dimensions of the two bodies are insignificant compared with the distance between the two, these conditions are fairly met. Putting values for P and a from Table 8.1, we obtain the mass of the Sun as 1.99×10^{30} kg.

The mass of a planet can similarly be determined from the sidereal periods of small orbiting satellites. For planets without any natural satellite, accurate mass determination requires orbital fly-byes of man-made spacecraft. Thus, the mass of Mars was accurately known long before those of Venus and Mercury.

8.11 The Two-Body Problem in the Solar System

The second condition of Keplerian motion mentioned in Sec. 8.1 was that the mass of the larger body was infinitely greater than that of the smaller body ($M \gg m$). For Keplerian motion of the planets, asteroids and cometary objects, this condition is, by and large, met. The greatest departure occurs in the case of Jupiter's motion around the Sun, where the **mass ratio** (also called the **reciprocal mass**) $M/m \approx 1,048$ (The New Solar System, 1999). Here M is the mass of the Sun and m stands for the mass of the Jovian system (Jupiter plus its satellites). Even smaller mass ratios are found in the planet–satellite systems. The Earth–Moon system has a mass raio of 81:1, whereas the Pluto–Charon system's mass ratio is only about 8:1 (from The New Solar System, 1999). Thus, until the demotion of Pluto from its planetary status in 2006, Pluto and Charon most closely resembled a **double planet** system.

Outside our solar system, the two-body problem is exemplified by the **binary star** system, where two stars revolve around their mutual center of mass. The more massive star is called the **primary star**, whereas the less massive star is called the **secondary star** or the **companion star** [cf. Van de Kamp (1964)].

8.12 Estimating the Mass of the Moon

The mass of the Moon has now been accurately determined from the period of man-made satellites placed in orbit around it (cf. Sec. 8.10). However, prior to space missions, a cruder method was used [cf. Cook (1980)]. It is based on the fact that the Earth and the Moon are a two-body system, whose center of mass moved in a nearly circular orbit around the Sun (Fig. 8.4). If the diameters of the orbits of the Earth and the Moon around their common center of mass are δ and d, respectively, then

$$\frac{\delta}{d} = \frac{m}{M}. \qquad (8.26)$$

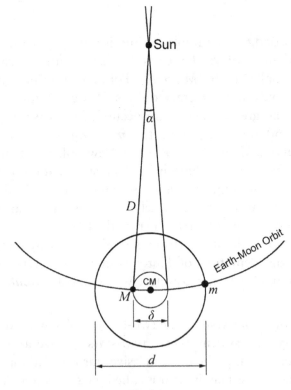

Fig. 8.4

If D is the distance of the Earth–Moon system from the Sun, the angle subtended by the Earth's orbit around the Earth–Moon center of mass at the Sun is

$$\alpha = \frac{\delta}{D}. \tag{8.27}$$

As seen from the extreme positions of the Earth's orbit around the center of mass, the Sun's location is shifted by this small angle against the distant star field in the celestial sphere (see Sec. 10.3). From the measured shift, the mass of the Moon follows from Eqs. (8.26) and (8.27):

$$m = \frac{\alpha M D}{d}. \tag{8.28}$$

It can be stated that the parallax of the Sun enables the determination of the mass of the Moon.

8.13 Determining the Masses of Venus and Mercury

Mercury and Venus are the only planets in the solar system which do not possess natural satellites and therefore their masses were not accurately known. Sending space probes to these planets have furnished accurate determination of their masses. For a spacecraft journey from the Earth to a destination planet, three phases of gravitational regime are encounted [cf. Bate *et al.* (1971)]: (1) In the first 'geocentric phase', the spacecraft is under the domination of the Earth's gravitational field. (2) Upon escaping the gravitational field of the Earth, the spacecraft enters the long 'heliocentric phase', where it is under the domain of the gravitational field of the Sun. (3) In the final phase, the spacecraft enters the gravitational field of the destination planet. In this last phase, the speed of the satellite is governed by the Equation of Energy:

$$v^2 = GM \left(\frac{2}{r} - \frac{1}{a} \right). \tag{2.76}$$

In this equation, M refers to the mass of the destination planet, a and r relate to the new orbit of the spacecraft around the destination planet. The speed of the spacecraft is determined from the Doppler shift of radio waves emitted from the spacecraft. Thus, the mass of the planet M can be determined. The masses of Venus and Mercury have been determined this way [cf. Cook (1980)].

8.14 The Three-Body Problem in the Solar System

An interesting special case of the three-body problem with applications in the solar system is the **restricted three-body problem**, in which two bodies are placed in circular orbits around their center of mass and the third body of negligible mass orbits at a fixed position relative to the two larger masses. This celebrated problem was solved by Lagrange in 1772. The mathematics of the solution is quite complicated and will not be given here. Extensive treatments are found in Moulton (1970) and Danby (1988). A more simplified solution is given in Van de Kamp (1964). We only give the results of this problem.

Figure 8.5 is the schematic diagram of a very special case of the restricted three-body problem which can be applied in the solar system. Masses M and m are placed in mutual Keplerian orbits. Let us assume that $M \gg m$. Then the center of mass is practically at M. Let us further assume that mass m revolves in a circular orbit of radius a around M. Next, the third mass $\mu (m \gg \mu)$ is brought into the plane of the circular orbit. Then, Lagrange's analysis yields five equilibrium points for μ, which are called the **Lagrangian points** or **libration points**. Three of them, marked L_1, L_2 and L_3 in Fig. 8.5 lie on the straight line joining M and m in accordance with Euler's colinear solutions of the three-body problem (Sec. 6.16). The rest, L_4 and L_5, are at the corners of equilateral triangles on either side of M and m as permitted by Lagrange's equilateral triangle solution of the three-body problem (Sec. 6.16). Denoting the distances of L_1,

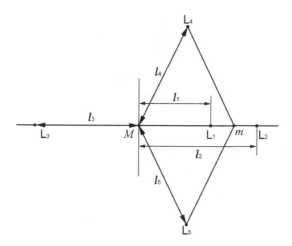

Fig. 8.5

L_2 and L_3 from the center of mass (and thus from M) as l_1, l_2 and l_3, respectively, then we have [from Van de Kamp (1964)]

$$l_1 \approx a\left(1 - \sqrt[3]{\frac{m}{3M}}\right),\qquad\qquad (8.29)$$

$$l_2 \approx a\left(1 + \sqrt[3]{\frac{m}{3M}}\right),\qquad\qquad (8.30)$$

and

$$l_3 \approx a.\qquad\qquad (8.31)$$

The distances l_4 and l_5 of L_4 and L_5 from M (and m) are obviously

$$l_4 = a,\qquad\qquad (8.32)$$

and

$$l_5 = a.\qquad\qquad (8.33)$$

L_1, L_2 and L_3 are unstable equilibrium points, whereas L_4 and L_5 represent stable equilibrium points. The mass μ placed at the first three points will slowly drift away, while they will stay at the last two Lagrangian points.

In the solar system, a large number of asteroids (several hundreds as of last count) have been found to occupy the L_4 and L_5 Lagrange points of Jupiter's orbit around the Sun. They are termed the ***Trojan asteroids***. A few Trojans have also been reported on Mars' orbit. No such asteroids have been found on the Earth's orbit so far, but only dust accumulation has.

For the Earth's orbit, the L_1 point is approximately 1.5 million kilometers from the Earth in the direction of the Sun. This distance is about 3.894 times the radius of the Moon's orbit. Even though it is an unstable point, L_1 is a convenient location for a spacecraft to observe the Sun continuously. The ISEE-3 spacecraft was placed there and carried out continuous observations of the Sun for several years. More recently, the SOHO spacecraft was positioned there for continuous monitoring of the Sun. Since the spacecrafts naturally drift away from unstable points, periodic corrections have to be carried out to keep them there.

The other Lagrangian points have also been targeted for spacecraft location in the future. For example, L_2 lies at the night side of the Earth and represents an ideal place to observe the greater universe. The European Space Agency has plans to place a spacecraft there in the near future. L_4 and L_5 are also being considered for possible sites for future space stations.

Exercises

8.1. Verify Eqs. (8.2) to (8.7).

8.2. The results of Secs. 8.2 and 8.3 are valid only because of the inverse square variation of the gravitational force. Discuss the similarity of the results with Gauss's Theorem in Electrostatics.

8.3. Show that the gravitational force on a mass particle anywhere inside a uniform hollow sphere is zero.

8.4. A mass particle is dropped into a straight frictionless tunnel through the center of the Earth. Show that the resulting motion is simple harmonic in nature. Obtain the period of oscillations. Compare this period with that of a satellite skimming the surface of the Earth in the absence of air resistance. Discuss the relevance of the results to the conservative nature of the gravitational force.

8.5. Using the data from Table 8.1, obtain the values of the following:

(a) The velocity of a satellite skimming the surface of the Earth without air drag [from Eq. (8.13)].
(b) The period of the satellite in (a) [from Eq. (8.15)].
(c) The equatorial velocity of escape from the surface of the Earth [from Eq. (8.18)].

8.6. (a) Obtain the velocity of a location on the Earth's equator due to the rotation of the Earth [from Eq. (8.14)].

(b) Obtain the velocity of a location on the Earth's surface at a latitude of 45° due to the Earth's rotation.

8.7. Obtain the time of travel of a spacecraft to the Moon [from Eq. (8.21)].

8.8. Obtain the time of travel of a spacecraft to Mars using the Hohmann transfer orbit [from Eq. (8.24)].

8.9. (a) Obtain the time of travel of a spacecraft to Venus using the Hohmann transfer orbit (Ans.: 146 days).

(b) Obtain the time of travel of a spacecraft to Mercury using the Hohmann transfer orbit (Ans.: 105 days).

(c) Discuss the results (a) and (b) from a physical standpoint.

8.10. (a) Find the location of the center of mass of the Earth–Moon system. Note that it is still inside the Earth. Due to tidal effect, the Moon is slowly receding from the Earth, and one day, the center of mass will lie outside the Earth. At that time, we will have a double planet in accordance with the definition of some astronomers.

(b) Find the location of the center of mass of the Pluto–Charon system. Note that it is entirely outside Pluto, which renders the system a double planet according to some astronomers. Note also that Pluto was demoted from its planetary status by the International Astronomers Association in August 2006.

8.11. (a) Obtain the locations of the unstable Lagrange points L_1, L_2 and L_3 of the Sun–Earth system [for an elementary derivation, see Van de Kamp (1964)].

(b) Obtain the location of the stable Langrange points L_4 and L_5 of the Sun–Earth system [see Van de Kamp (1964), for an elementary derivation].

Planetary Motion in Three-Dimensional Space

9.1 Keplerian Motion in Three-Dimensional Space

Owing to the fact that the motion of a planet lies in a plane, we had, hitherto, described this motion in two-dimensional Cartesian and polar coordinates. But space is three-dimensional and the luxury of restricting the motion in two dimensions is not always there. Thus we need to view the planetary problem in three-dimensional space also.

9.2 Newton's Law of Gravitation in Three Dimensions

As usual, we treat both the Sun and the planet as point particles and assume that the Sun is infinitely more massive than the planet ($M \gg m$). In Fig. 9.1, the Sun is placed at the origin of a three-dimensional coordinate system. The position vector of the planet is

$$\vec{r} = r\hat{r}. \tag{9.1}$$

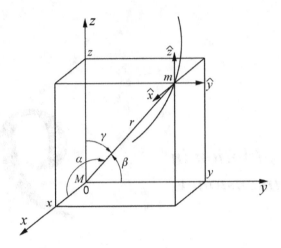

Fig. 9.1

Then the velocity and the acceleration vectors are given by

$$\vec{v} = \frac{d\vec{r}}{dt} = \dot{\vec{r}}, \tag{9.2}$$

and

$$\vec{a} = \frac{d\vec{v}}{dt} = \frac{d^2\vec{r}}{dt^2} = \dot{\vec{v}} = \ddot{\vec{r}}. \tag{9.3}$$

By Newton's law, the gravitational force of the Sun on the planet is given by

$$\vec{f} = -\frac{GMm}{r^2}\hat{r} = -\frac{GMm}{r^3}\vec{r}. \tag{9.4}$$

On account of Newton's second law of motion, we have

$$m\ddot{\vec{r}} = -\frac{GMm}{r^3}\vec{r}. \tag{9.5}$$

Thus, the equation of motion of the planet in three-dimensional space takes the form

$$\ddot{\vec{r}} + \frac{GM}{r^3}\vec{r} = \vec{0}. \tag{9.6}$$

9.3 Conservation of Angular Momentum in Three Dimensions

By taking the cross-product of $\vec{r}$ with Eq. (9.6), we get

$$\vec{r} \times \ddot{\vec{r}} + \frac{GM}{r^3} \vec{r} \times \vec{r} = \vec{0}. \tag{9.7}$$

Since the second term on the right-hand side is zero, we have

$$\vec{r} \times \ddot{\vec{r}} = \vec{0}. \tag{9.8}$$

Now

$$\frac{d}{dt}(\vec{r} \times \dot{\vec{r}}) = \dot{\vec{r}} \times \dot{\vec{r}} + \vec{r} \times \ddot{\vec{r}} = \vec{r} \times \ddot{\vec{r}}. \tag{9.9}$$

Thus, from Eq. (9.8),

$$\frac{d}{dt}(\vec{r} \times \dot{\vec{r}}) = \vec{0}. \tag{9.10}$$

Multiplying Eq. (9.10) by m, we obtain the law of conservation of angular momentum:

$$\vec{l} = \vec{r} \times \vec{p} = const. \tag{9.11}$$

Alternatively, from Eq. (9.4), the gravitational torque on the planet is zero (cf. Sec. 4.2):

$$\vec{\tau} = \vec{r} \times \vec{f} = -\frac{GMm}{r^3} \vec{r} \times \vec{r} = \vec{0}. \tag{9.12}$$

But

$$\vec{\tau} = \frac{d\vec{l}}{dt}. \tag{4.5}$$

Thus, the angular momentum for orbital motion $\vec{l}$ is a constant.

Some authors [e.g. Bate *et al.* (1971)] define the angular momentum per unit mass as the *specific angular momentum*

$$\vec{k} = \frac{\vec{l}}{m} = \vec{r} \times \vec{v} = \vec{r} \times \dot{\vec{r}}. \tag{9.13}$$

9.4 Conservation of Total Energy in Three Dimensions

Taking the dot-product of $\dot{\vec{r}}$ with Eq. (9.6), we have

$$\dot{\vec{r}} \cdot \ddot{\vec{r}} + \frac{GM}{r^3} \dot{\vec{r}} \cdot \vec{r} = 0, \tag{9.14}$$

or

$$\vec{v} \cdot \dot{\vec{v}} + \frac{GM}{r^3} \dot{\vec{r}} \cdot \vec{r} = 0. \tag{9.15}$$

As an exercise, one can show that

$$\vec{v} \cdot \dot{\vec{v}} = v\dot{v}. \tag{9.16}$$

Similarly,

$$\dot{\vec{r}} \cdot \vec{r} = r\dot{r}. \tag{9.17}$$

Thus Eq. (9.15) assumes the form

$$v\dot{v} + \frac{GM}{r^2}\dot{r} = 0. \tag{9.18}$$

We observe that

$$\frac{d}{dt}\left(\frac{1}{2}v^2\right) = v\dot{v}. \tag{9.19}$$

Also,

$$\frac{d}{dt}\left(-\frac{GM}{r}\right) = \frac{GM}{r^2}\dot{r}. \tag{9.20}$$

Hence Eq. (9.18) becomes

$$\frac{d}{dt}\left(\frac{1}{2}v^2 - \frac{GM}{r}\right) = 0. \tag{9.21}$$

By multiplying Eq. (9.21) by m, we get the conservation of total energy of the planet:

$$E = \frac{1}{2}mv^2 - \frac{GMm}{r} = const. \tag{9.22}$$

The total energy divided by the mass is also called the ***specific total energy*** [e.g. Bate *et al.* (1971)].

9.5 Constancy of the Runge–Lenz Vector in Three Dimensions

Taking the cross-product of Eq. (9.6) with $\vec{k}$, we have

$$\ddot{\vec{r}} \times \vec{k} = -\frac{GM}{r^3}\vec{r} \times \vec{k} = \frac{GM}{r^3}\vec{k} \times \vec{r}. \tag{9.23}$$

Now

$$\frac{d}{dt}(\dot{\vec{r}} \times \vec{k}) = \ddot{\vec{r}} \times \vec{k} + \dot{\vec{r}} \times \dot{\vec{k}}. \tag{9.24}$$

Since $\dot{\vec{k}} = \vec{0}$ by the conservation of angular momentum, the left-hand side of Eq. (9.23) is written as

$$\ddot{\vec{r}} \times \dot{\vec{k}} = \frac{d}{dt}(\dot{\vec{r}} \times \vec{k}). \tag{9.25}$$

We now show that the right-hand side of Eq. (9.23) is also equal to a time derivative. We have

$$\frac{GM}{r^3}\vec{k} \times \vec{r} = \frac{GM}{r^3}(\vec{r} \times \dot{\vec{r}}) \times \vec{r} = -\frac{GM}{r^3}\vec{r} \times (\vec{r} \times \dot{\vec{r}}). \tag{9.26}$$

Expanding the vector triple product, we get

$$\frac{GM}{r^3}\vec{k} \times \vec{r} = -\frac{GM}{r^3}(\vec{r} \cdot \dot{\vec{r}})\vec{r} + \frac{GM}{r^3}r^2\dot{\vec{r}}. \tag{9.27}$$

By virtue of Eq. (9.17),

$$\frac{GM}{r^3}\vec{k} \times \vec{r} = -\frac{GM}{r^2}\dot{r}\vec{r} + \frac{GM}{r}\dot{\vec{r}}. \tag{9.28}$$

Next,

$$\frac{d}{dt}\left[GM\left(\frac{\vec{r}}{r}\right)\right] = GM\frac{d}{dt}\left(\frac{1}{r}\vec{r}\right) = -\frac{GM}{r^2}\dot{r}\vec{r} + \frac{GM}{r}\dot{\vec{r}}. \tag{9.29}$$

Thus, Eq. (9.23) is finally transformed into

$$\frac{d}{dt}(\dot{\vec{r}} \times \vec{k}) = \frac{d}{dt}\left(\frac{GM}{r}\vec{r}\right),$$
(9.30)

which is equivalent to

$$\frac{d}{dt}(\vec{p} \times \vec{l} - GMm^2\hat{r}) = \vec{0}.$$
(9.31)

Equation (9.31) states the conservation of the Runge–Lenz vector $\vec{R}$ of Eq. (4.84):

$$\vec{R} = \vec{p} \times \vec{l} - GMm^2\hat{r} = const.$$
(9.32)

9.6 The Equation of Orbit in Three Dimensions

By taking the dot-product of Eq. (9.32) with $\vec{r}$, we have

$$\vec{R} \cdot \vec{r} = (\vec{p} \times \vec{l}) \cdot \vec{r} - \frac{GMm^2}{r}\vec{r} \cdot \vec{r}.$$
(9.33)

Let θ be the angle between the invariant direction of $\vec{R}$ and $\vec{r}$. By cyclic permutation of the scalar triple product on the right-hand side, we get

$$Rr\cos\theta = (\vec{r} \times \vec{p}) \cdot \vec{l} - GMm^2r = l^2 - GMm^2r.$$
(9.34)

Rearranging algebraically,

$$r = \frac{\dfrac{l^2}{GMm^2}}{1 + \dfrac{R}{GMm^2}\cos\theta},$$
(9.35)

which is the equation of a conic [cf. Eq. (4.138)] of the semi-latus rectum,

$$p = \frac{l^2}{GMm^2},$$
(9.36)

and eccentricity,

$$e = \frac{R}{GMm^2}.$$ (9.37)

9.7 The Two-Body Problem in Three Dimensions

Consider the two-body problem with masses m_1 and m_2 whose position vectors are $\vec{r}_1$ and $\vec{r}_2$ in an inertial frame of reference (Fig. 9.2). Let $\vec{r}$ be the position vector of m_2 relative to m_1. Then, by vector addition, we have

$$\vec{r}_1 + \vec{r} = \vec{r}_2.$$ (9.38)

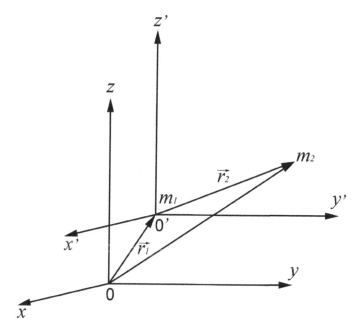

Fig. 9.2

The equation of motion of m_2 is written down from Newton's second law of motion and his law of gravitation:

$$m_2\ddot{\vec{r}}_2 = -\frac{Gm_1m_2}{r^3}\vec{r}, \tag{9.39}$$

or

$$\ddot{\vec{r}}_2 = -\frac{Gm_1}{r^3}\vec{r}. \tag{9.40}$$

The equation of motion of m_1 as given by Newton's third law of motion is

$$m_1\ddot{\vec{r}}_1 = \frac{Gm_2m_1}{r^3}\vec{r}, \tag{9.41}$$

or

$$\ddot{\vec{r}}_1 = \frac{Gm_2}{r^3}\vec{r}. \tag{9.42}$$

Differentiating Eq. (9.38) twice with respect to time gives

$$\ddot{\vec{r}} = \ddot{\vec{r}}_2 - \ddot{\vec{r}}_1. \tag{9.43}$$

Also, from Eqs. (9.40) and (9.42), we have

$$\ddot{\vec{r}}_2 - \ddot{\vec{r}}_1 = -\frac{G(m_1 + m_2)}{r^3}\vec{r}. \tag{9.44}$$

Finally, from Eqs. (9.43) and (9.44):

$$\ddot{\vec{r}} = -\frac{G(m_1 + m_2)}{r^3}\vec{r}. \tag{9.45}$$

Thus, the equation of motion in the relative coordinate system with m_1 at the origin is of the same form as that of the one-body motion [Eq. (9.5)] with the total mass of the system replacing m_1.

9.8 The Planetary Problem in Rectangular Coordinates

In a rectangular coordinate system (x, y, z), the unit vectors $\hat{x}, \hat{y}$ and $\hat{z}$ are all constants (cf. Fig. 9.1). Furthermore, by the three-dimensional Pythagorean Theorem,

$$x^2 + y^2 + z^2 = r^2. \tag{9.46}$$

Hence the three components of Eq. (9.6) are as follows:

$$\ddot{x} = -\frac{GMx}{\left(x^2 + y^2 + z^2\right)^{3/2}}, \tag{9.47}$$

$$\ddot{y} = -\frac{GMy}{\left(x^2 + y^2 + z^2\right)^{3/2}}, \tag{9.48}$$

and

$$\ddot{z} = -\frac{GMz}{\left(x^2 + y^2 + z^2\right)^{3/2}}. \tag{9.49}$$

Equations (9.47)–(9.49) are a set of nonlinear coupled differential equations and there is no general way to uncouple them. In the special case of motion in a circle, r is constant and x, y, z are all period functions of time.

One can still manage to obtain some results in the general case. Equations (9.47)–(9.49) readily give

$$y\ddot{z} - z\ddot{y} = 0, \tag{9.50}$$
$$z\ddot{x} - x\ddot{z} = 0, \tag{9.51}$$

and

$$x\ddot{y} - y\ddot{x} = 0. \tag{9.52}$$

Equations (9.50)–(9.52) can be integrated over t, giving

$$y\dot{z} - z\dot{y} = c_1, \tag{9.53}$$
$$z\dot{x} - x\dot{z} = c_2, \tag{9.54}$$

and

$$x\dot{y} - y\dot{x} = c_3, \tag{9.55}$$

where c_1, c_2 and c_3 are arbitrary constants of integration. From Eqs. (9.53), (9.54) and (9.55), we get

$$c_1 x + c_2 y + c_3 z = 0, \qquad (9.56)$$

which is the equation of a plane passing through the origin (i.e. the attracting center) [cf. Brouwer and Clemence (1961)]. Thus, the orbit of a planet lies on a plane, a result obtained for the general case of the central force motion.

Equations (9.53)–(9.55) also betray the constancy of the angular momentum components in planetary motion. By Eq. (9.11), we have

$$\vec{l} = \vec{r} \times \vec{p} = \vec{r} \times m\dot{\vec{r}} = m \begin{vmatrix} \hat{x} & \hat{y} & \hat{z} \\ x & y & z \\ \dot{x} & \dot{y} & \dot{z} \end{vmatrix}. \qquad (9.57)$$

Thus

$$l_x = m(y\dot{z} - z\dot{y}) = mc_1 = const., \qquad (9.58)$$
$$l_y = m(z\dot{x} - x\dot{z}) = mc_2 = const., \qquad (9.59)$$

and

$$l_z = m(x\dot{y} - y\dot{x}) = mc_3 = const. \qquad (9.60)$$

From Eqs. (9.47)–(9.49), we get

$$\dot{x}\ddot{x} + \dot{y}\ddot{y} + \dot{z}\ddot{z} = -\frac{GM}{r^3}(x\dot{x} + y\dot{y} + z\dot{z}), \qquad (9.61)$$

or

$$\frac{1}{2}\frac{d}{dt}(\dot{x}^2 + \dot{y}^2 + \dot{z}^2) = -\frac{GM}{r^3}\frac{d}{dt}(x^2 + y^2 + z^2). \qquad (9.62)$$

Hence

$$\frac{1}{2}\frac{d}{dt}(v^2) = -\frac{1}{2}\frac{GM}{r^3}\frac{d}{dt}(r^2) = -\frac{GM}{r^2}\frac{dr}{dt} = GM\frac{d}{dt}\left(\frac{1}{r}\right). \qquad (9.63)$$

Integrating over t and multiplying by m, one obtains the equation of energy:

$$\frac{1}{2}mv^2 - \frac{GMm}{r} = E. \tag{9.64}$$

If α, β and γ are the angles of the position vector from the x-, y- and z-axes, respectively [cf. Fig. 9.1], then Eqs. (9.47)–(9.49) can be re-written as

$$\ddot{x} = -\frac{GM}{r^2}\cos\alpha, \tag{9.65}$$

$$\ddot{y} = -\frac{GM}{r^2}\cos\beta, \tag{9.66}$$

and

$$\ddot{z} = -\frac{GM}{r^2}\cos\gamma, \tag{9.67}$$

where $\cos\alpha$, $\cos\beta$ and $\cos\gamma$ are the **direction cosines**. Since they are related by the condition

$$\cos^2\alpha + \cos^2\beta + \cos^2\gamma = 1, \tag{9.68}$$

the motions in the three directions are not all independent. This is consistent with the fact that the motion of a planet lies on one plane.

9.9 The Planetary Problem in Spherical Coordinates

The spherical coordinates (r, θ, ϕ) are an orthogonal coordinate system having one linear coordinate and two angular coordinates, which are the radial coordinate $r(0 \leq r \leq \infty)$, the zenith angle $\theta(0 \leq \theta \leq \pi)$, and the azimuth angle $\phi(0 \leq \phi \leq 2\pi)$ (cf. Fig. 9.3). The transformation from spherical to rectangular coordinates are given by

$$x = r\sin\theta\cos\phi, \tag{9.69}$$

$$y = r\sin\theta\sin\phi, \tag{9.70}$$

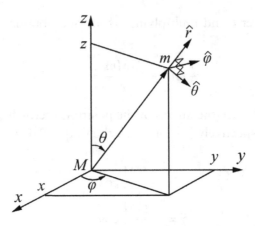

Fig. 9.3

and

$$z = r\cos\theta.$$

(9.71)

The inverse transformation relations are given by

$$r = \sqrt{x^2 + y^2 + z^2},$$

(9.72)

$$\theta = \cos^{-1} \frac{z}{\sqrt{x^2+y^2+z^2}},$$

(9.73)

and

$$\phi = \tan^{-1} \frac{y}{x}.$$

(9.74)

The transformation relations between the unit vectors in the two coordinate systems are given by [cf. Arfken and Weber (2005)]

$$\hat{x} = \hat{r}\sin\theta\cos\phi + \hat{\theta}\cos\theta\cos\phi - \hat{\phi}\sin\phi,$$

(9.75)

$$\hat{y} = \hat{r}\sin\theta\sin\phi + \hat{\theta}\cos\theta\sin\phi + \hat{\phi}\cos\phi,$$

(9.76)

and

$$\hat{z} = \hat{r}\cos\theta - \hat{\theta}\sin\theta,$$

(9.77)

whereas the inverse transformations are given by [cf. Arfken and Weber (2005)]

$$\hat{r} = \hat{x} \sin\theta \cos\phi + \hat{y} \sin\theta \sin\phi + \hat{z} \cos\theta, \tag{9.78}$$

$$\hat{\theta} = \hat{x} \cos\theta \cos\phi + \hat{y} \cos\theta \sin\phi - \hat{z} \sin\theta, \tag{9.79}$$

and

$$\hat{\phi} = -\hat{x} \sin\phi + \hat{y} \cos\phi. \tag{9.80}$$

The unit vectors $\hat{r}, \hat{\theta}$ and $\hat{\phi}$ are all constants in the radial direction, but not in the transverse directions. We have [cf. Arfken and Weber (2005)]:

$$\frac{\partial \hat{r}}{\partial r} = 0, \tag{9.81}$$

$$\frac{\partial \hat{\theta}}{\partial r} = 0, \tag{9.82}$$

$$\frac{\partial \hat{\phi}}{\partial r} = 0, \tag{9.83}$$

$$\frac{\partial \hat{r}}{\partial \theta} = \hat{\theta}, \tag{9.84}$$

$$\frac{\partial \hat{\theta}}{\partial \theta} = -\hat{r}, \tag{9.85}$$

$$\frac{\partial \hat{\phi}}{\partial \phi} = 0, \tag{9.86}$$

$$\frac{\partial \hat{r}}{\partial \phi} = \sin\theta \hat{\phi}, \tag{9.87}$$

$$\frac{\partial \hat{\theta}}{\partial \phi} = \cos\theta \hat{\phi}, \tag{9.88}$$

and

$$\frac{\partial \hat{\phi}}{\partial \phi} = -\hat{r} \sin\theta - \hat{\theta} \cos\theta. \tag{9.89}$$

The Lagrangian for the planetary problem in spherical coordinates is written as

$$L = \frac{1}{2}m\dot{r}^2 + \frac{1}{2}mr^2\dot{\theta}^2 + \frac{1}{2}mr^2\sin^2\theta\dot{\phi}^2 + \frac{GMm}{r}. \qquad (9.90)$$

Lagrange's equations of motion in spherical coordinates are the following:

$$m\ddot{r} - mr\dot{\theta}^2 - mr\sin^2\theta\dot{\phi}^2 + \frac{GMm}{r^2} = 0, \qquad (9.91)$$

$$\frac{d}{dt}(mr^2\dot{\theta}) - mr^2\sin\theta\cos\theta\dot{\phi}^2 = 0, \qquad (9.92)$$

and

$$\frac{d}{dt}(mr^2\sin^2\theta\dot{\phi}) = 0. \qquad (9.93)$$

The generalized momenta are given by

$$p_r = \frac{\partial L}{\partial \dot{r}} = m\dot{r}, \qquad (9.94)$$

$$p_\theta = \frac{\partial L}{\partial \dot{\theta}} = mr^2\dot{\theta}, \qquad (9.95)$$

and

$$p_\phi = \frac{\partial L}{\partial \dot{\phi}} = mr^2\sin^2\theta\dot{\phi}. \qquad (9.96)$$

Whereas p_r represents the linear momentum of the planet in the radial direction, p_θ is the angular momentum about an axis perpendicular to the meridional plane (reckoned positive if southwards) and p_ϕ gives the angular momentum about the z-axis (reckoned positive if eastwards). Since the Lagrangian does not contain ϕ explicitly, the latter is an ignorable coordinate and p_ϕ is a constant of motion [Eqs. (9.93) and (9.96)].

The Hamiltonian is constructed as follows:

$$H = \sum_i p_i\dot{q}_i - L = \frac{1}{2}m\dot{r}^2 + \frac{1}{2}mr^2\dot{\theta}^2 + \frac{1}{2}mr^2\sin^2\theta\dot{\phi}^2 - \frac{GMm}{r} = E. \qquad (9.97)$$

The Hamiltonian is equal to the total energy since the system is conservative and the kinetic energy is a homogeneous quadratic function of the generalized velocities. In terms of generalized coordinates and generalized momenta, the Hamiltonian is expressed by convention as

$$H(r, \theta, \phi; p_r, p_\theta, p_\phi) = \frac{p_r^2}{2m} + \frac{p_\theta^2}{2mr^2} + \frac{p_\phi^2}{2mr^2 \sin^2 \theta} - \frac{GMm}{r}.$$
$$(9.98)$$

One set of Hamilton's equations of motion give the definitions of the generalized momenta:

$$\dot{r} = \frac{\partial H}{\partial p_r} = \frac{p_r}{m}, \tag{9.99}$$

$$\dot{\theta} = \frac{\partial H}{\partial p_\theta} = \frac{p_\theta}{mr^2}, \tag{9.100}$$

and

$$\dot{\phi} = \frac{\partial H}{\partial p_\phi} = \frac{p_\phi}{mr^2 \sin^2 \theta}. \tag{9.101}$$

The second set of Hamilton's equations give:

$$\dot{p}_r = -\frac{\partial H}{\partial r} = \frac{p_\theta^2}{mr^3} + \frac{p_\phi^2}{mr^3 \sin^2 \theta} - \frac{GMm}{r^2}, \tag{9.102}$$

$$\dot{p}_\theta = -\frac{\partial H}{\partial \theta} = \frac{p_\phi^2 \cos \theta}{mr^2 \sin^3 \theta}, \tag{9.103}$$

and

$$\dot{p}_\phi = -\frac{\partial H}{\partial \phi} = 0. \tag{9.104}$$

Equations (9.102)–(9.104) are equivalent to Lagrange's equations of motion (9.91)–(9.93), respectively. However, unless simplifications are made, the equations remain intrinsically coupled.

Further advances in planetary motion have been made using the *Hamilton–Jacobi Theory* of classical mechanics, particularly in the three-body problem. In fact, the Hamilton–Jacobi formulation was borne out of the study of planetary motion. Unfortunately, the level of analysis is a notch higher than that of this book.

Exercises

9.1. Verify Eqs. (9.16) and (9.17).

9.2. Verify Eqs. (9.19) and (9.20).

9.3. Verify Eq. (9.27).

9.4. Verify Eq. (9.29).

9.5. Verify Eqs. (9.34) and (9.35).

9.6. Verify Eqs. (9.53) to (9.55).

9.7. Verify Eqs. (9.62) to (9.64).

9.8. Derive Eqs. (9.75) to (9.77).

9.9. Derive Eqs. (9.78) to (9.80).

9.10. Show that $\hat{\theta} = \frac{\partial \hat{r}}{\partial \theta}$.

9.11. Show that $\hat{\phi} = \frac{1}{\sin\theta} \frac{\partial \hat{r}}{\partial \phi}$.

9.12. Obtain Eq. (9.90).

9.13. Obtain Eqs. (9.97) and (9.98).

9.14. Obtain Eqs. (9.99) to (9.104).

Motion of Artificial Earth Satellites

10.1 Artificial Earth Satellites

The launch of Sputnik 1 on October 4, 1957 ushered in the space age which continues today. Thousands of satellites of various kinds have been put into orbit by various nations. Hundreds of them serve our everyday lives at any time. We are truly in a space age today.

As previously asserted, the orbit of a satellite is Keplerian with the center of the Earth at one focus. Since any central force orbit lies on a plane, we had, heretofore, the luxury of treating the problem in two-dimensional space. The location of the observer was immaterial. In the case of the artificial satellites, however, this luxury is no longer affordable. Since the observer is on the surface of the Earth, the orbits of satellites must be monitored in the three-dimensional space.

10.2 Frames of Reference

Any frame of reference (a coordinate system) in which Newton's laws of motion are valid is called an **inertial frame of reference**. In such a frame, we can write down the equations of motion, which are force equations given by Newton's second law of motion, and solve a dynamical problem. Any frame of reference moving with a constant velocity relative to an inertial frame of reference, is also another inertial frame. This is because the equations of motion or the force equations are the same in the two frames, even though their integrals (viz., the velocities and coordinates) are different. If the second frame of reference has an acceleration relative to the inertial frame, then that frame of reference is no longer an inertial frame. Similarly, a frame of reference rotating relative to an inertial frame, is not an inertial frame. In such a frame, the centrifugal and Coriolis forces always come into play, which are not prescribed by Newton's second law of motion.

Finding an inertial frame of reference depends on the problem at hand. For an experiment in the laboratory, the walls of the laboratory may be taken to be the inertial frame. For a localized event on the Earth's surface, the surface of the Earth may be the reference frame of choice. For the motion of planets around the Sun, a heliocentric frame of reference with an axis pointing towards the axis of rotation of the Sun can be chosen. Likewise, for the motion of a satellite around the Earth, a **geocentric-equatorial system of coordinates** is chosen [cf. Bate *et al.* (1971)]. In such a system (Fig. 10.1), the center of the Earth is the origin. The invariant diection given by the rotational angular momentum of the Earth along its axis of rotation points towards the **pole star** (currently Polaris or the "north star"). This direction is defined as the positive z-axis. The x- and y-axis lie on the equatorial plane of the Earth. The x-axis points towards the vernal equinox direction (which the Sun crosses most frequently on March 20) and the y-axis is perpendicular to it, such that (x, y, z) form a right-handed system. Such a coordinate system does not rotate relative to the

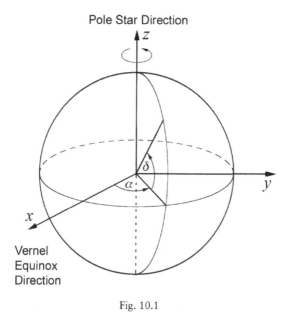

Fig. 10.1

fixed distant star-field, and is therefore, a legitimate inertial frame of reference.

10.3 The Celestial Sphere, Right Ascension and Declination

In the geocentric-equatorial system of coordinates, the distant star field has the same configuration at all times and the stars appear to be "fixed" on sphere of "infinite radius" called the *celestial sphere.* The celestial sphere serves as a convenient background to which the locations of relatively nearby objects such as the Sun, the Moon, the planets and the satellites, are referred.

The location of a satellite on the celestial sphere is determined by two angular coordinates akin to the latitude and longitude of a place on the surface of the globe. The *right ascension* is the angle measured on the equatorial plane from the vernal equinox position (angle α in Fig. 10.1). This is similar to the azimuth angle in spherical coordinates and longitude of a location on the globe. The *declination* is the angle measured from the equatorial plane at the center of the coordinate system (angle δ in Fig. 10.1). This is complement of the zenith angle in spherical coordinates from 90° or the latitude of a location on the globe.

10.4 The Orbital Elements of a Satellite

In order to unambiguously specify the orbit of a satellite in three-dimensional space and its location in that orbit, a total of six quantities are required. They are referred to as the classical *orbital elements* of the satellite. Three quantities are required to specify the orientation of the orbit in space; two quantities are required to specify the "size" and "shape" of the orbit; and finally, one quantity gives the location of the satellite at any instant on that orbit. The following are a set of commonly used orbital elements in the geocentric-equatorial coordinate system (Fig. 10.2):

Inclination. The inclination is the angle i of the orbital angular momentum vector of the satellite measured from the rotational axis of the Earth (Fig. 10.2). It is also the angle between the orbital plane of the satellite and the equatorial plane of the Earth. The orbit of the satellite intersects the equatorial plane of the Earth at two points called the *nodes*. The point of passage of the satellite from the southern hemisphere to the northern hemisphere is called the *ascending node*. The *descending node* marks the point where the satellite crosses the equatorial plane from the northern hemisphere. The inclination is equal to the angle between the orbital plane and the equatorial plane measured from the latter at the ascending node. Thus the value of the inclination lies between 0 and π (180°).

Semi-major axis. The semi-major axis of the orbit of the satellite a gives the "size" or "energy" of the orbit.

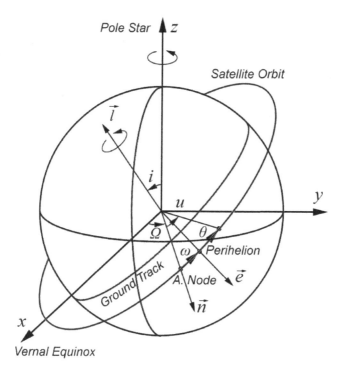

Fig. 10.2

Eccentricity. The eccentricity of the orbit of the satellite e gives the "shape" or "ellipticity" of the orbit.

Longitude of the ascending node. The longitude of the ascending node Ω (also called the ***right ascension of ascending node***) is the angle of the ascending node measured from the vernal equinox direction in the equatorial plane (Fig. 10.2). The range of Ω is between 0 and 2π (360°).

Argument of perigee. The argument of perigee ω is the angle of the perigee of the satellite measured from the ascending node in the orbital plane (Fig. 10.2). ω lies between 0 and π (180°).

True anomaly. The true anomaly of the satellite at any instant θ is the angle measured from the perigee in the equatorial plane (Fig. 10.2). The range of θ is between 0 and 2π (360°).

The above six orbital elements are by no means unique, and alternative representations are possible. For example, various representations of the orbital ellipse were given in Table 6.1. Also, the *height of perigee* h_P and the *height of apogee* h_A are often used. If r_0 is the reference radius of the Earth (conventionally taken as the equatorial radius, R_e of Table 8.1), then we have

$$r_0 + h_A = a(1 + e), \tag{10.1}$$

and

$$r_0 + h_P = a(1 - e). \tag{10.2}$$

By elimination, we obtain, from Eqs. (10.1) and (10.2):

$$a = r_0 + \frac{h_A + h_P}{2}, \tag{10.3}$$

and

$$e = \frac{h_A - h_P}{2a}. \tag{10.4}$$

From Fig. 10.2, it is clear that the argument of perigee ω and the true anomaly θ are amenable to scalar addition. Their sum is called the *argument of latitude u*:

$$u = \omega + \theta. \tag{10.5}$$

Interestingly, for a circular orbit ($e = 0$), both ω and θ are undefined, but u is still defined.

10.5 Three Fundamental Vectors

There exist three fundamental vectors which contain all the information on the orbital elements of a satellite. First, the specific angular momentum vector (Sec. 9.2) defines the orbital plane of the satellite to which it is perpendicular:

$$\vec{h} = \frac{1}{m}\vec{r} \times \vec{p} = \vec{r} \times \vec{v}. \tag{10.6}$$

Next, the eccentricity vector (Sec. 4.12), directed along the perihelion, defines orientation of the major axis:

$$\vec{e} = \frac{1}{GMm^2}\vec{R}, \tag{10.7}$$

with the Runge–Lenz vector,

$$\vec{R} = \vec{p} \times \vec{l} - GMm^2\hat{r}. \tag{10.8}$$

By expanding the triple vector product in $\vec{R}$, we can express the eccentricity vector in terms of the position and velocity vectors:

$$\vec{e} = \frac{1}{GM}(\vec{v} \times \vec{h}) - \frac{\vec{r}}{r} = \frac{1}{GM}\left[\left(v^2 - \frac{GM}{r}\right)\vec{r} - (\vec{r} \cdot \vec{v})\vec{v}\right]. \tag{10.9}$$

The third vector, called the **node vector**, defines the line of intersection of the orbital plane of the satellite and the equatorial plane of the Earth, towards the ascending node. In the geocentric–equatorial coordinates (x, y, z) with $\hat{z}$ towards the pole star and $\hat{x}$ towards the vernal equinox direction, the node vector is defined as

$$\vec{n} = \hat{z} \times \vec{h}. \tag{10.10}$$

10.6 Determination of Orbital Elements

Amongst the many essential applications of the radar are the tracking of artificial Earth satellites and spacecraft. A radar measurement gives both the position and velocity vectors of the target. The two independent vectors yield six independent components which are sufficient for determining all six orbital elements of a satellite.

Let the position and velocity vectors of the satellite in the geocentric-equatorial system be

$$\vec{r} = x\hat{x} + y\hat{y} + z\hat{z}, \qquad (10.11)$$

and

$$\vec{v} = v_x\hat{x} + v_y\hat{y} + v_z\hat{z}, \qquad (10.12)$$

with

$$r = \sqrt{x^2 + y^2 + z^2}, \qquad (10.13)$$

and

$$v = \sqrt{v_x^2 + v_y^2 + v_z^2}. \qquad (10.14)$$

The specific angular momentum vector and its magnitude are given by

$$\vec{h} = \begin{vmatrix} \hat{x} & \hat{y} & \hat{z} \\ x & y & z \\ v_x & v_y & v_z \end{vmatrix} = \left(yv_z - zv_y\right)\hat{x} + \left(zv_x - xv_z\right)\hat{y} + \left(xv_y - yv_x\right)\hat{z},$$

$$(10.15)$$

and

$$h = \sqrt{h_x^2 + h_y^2 + h_z^2}. \qquad (10.16)$$

The node vector and its magnitude are likewise

$$\vec{n} = \begin{vmatrix} \hat{x} & \hat{y} & \hat{z} \\ 0 & 0 & 1 \\ h_x & h_y & h_z \end{vmatrix} = -h_y\hat{x} + h_x\hat{y}, \qquad (10.17)$$

and

$$n = \sqrt{h_x^2 + h_y^2}. \tag{10.18}$$

The eccentricity vector can be calculated from either form of Eq. (10.9). We have

$$\vec{v} \times \vec{h} = \begin{vmatrix} \hat{x} & \hat{y} & \hat{z} \\ v_x & v_y & v_z \\ h_x & h_y & h_z \end{vmatrix}$$

$$= \left(v_y h_z - v_z h_y\right) \hat{x} + \left(v_z h_x - v_x h_z\right) \hat{y}$$
$$+ \left(v_x h_y - v_y h_x\right) \hat{z}. \tag{10.19}$$

Thus, from Eq. (10.9),

$$e_x = \frac{1}{GM} \left(v_y h_z - v_z h_y\right) - \frac{x}{r}, \tag{10.20}$$

$$e_y = \frac{1}{GM} \left(v_z h_x - v_x h_z\right) - \frac{y}{r}, \tag{10.21}$$

$$e_z = \frac{1}{GM} \left(v_x h_y - v_y h_x\right) - \frac{z}{r}, \tag{10.22}$$

and

$$e = \sqrt{e_x^2 + e_y^2 + e_z^2}. \tag{10.23}$$

Alternatively, from Eq. (10.9), we have

$$e_x = \frac{1}{GM} \left[\left(v^2 - \frac{GM}{r}\right) x - (\vec{r} \cdot \vec{v}) v_x \right], \tag{10.24}$$

$$e_y = \frac{1}{GM} \left[\left(v^2 - \frac{GM}{r}\right) y - (\vec{r} \cdot \vec{v}) v_y \right], \tag{10.25}$$

$$e_z = \frac{1}{GM} \left[\left(v^2 - \frac{GM}{r}\right) z - (\vec{r} \cdot \vec{v}) v_z \right], \tag{10.26}$$

with

$$\vec{r} \cdot \vec{v} = x v_x + y v_y + z v_z. \tag{10.27}$$

The orbital elements i and Ω are readily provided by the components of $\vec{h}$ and $\vec{n}$ along the z- and x-axis, respectively [cf. Bate *et al.* (1971)]. We have (vide Fig. 10.2)

$$\vec{h} \cdot \hat{z} = h_z = h \cos i, \tag{10.28}$$

and

$$\vec{n} \cdot \hat{x} = n_x = n \cos \Omega. \tag{10.29}$$

Thus,

$$i = \cos^{-1} \frac{h_z}{h}, \tag{10.30}$$

and

$$\Omega = \cos^{-1} \frac{n_x}{n}. \tag{10.31}$$

The elements ω, θ and u can be determined from the scalar products between $\vec{n}$, $\vec{e}$ and $\vec{r}$ [cf. Bate *et al.* (1971)]. We have (vide Fig. 10.2)

$$\vec{n} \cdot \vec{e} = ne \cos \omega, \tag{10.32}$$
$$\vec{e} \cdot \vec{r} = er \cos \theta, \tag{10.33}$$

and

$$\vec{n} \cdot \vec{r} = nr \cos u, \tag{10.34}$$

whence

$$\omega = \cos^{-1} \frac{\vec{n} \cdot \vec{e}}{ne}, \tag{10.35}$$

$$\theta = \cos^{-1} \frac{\vec{e} \cdot \vec{r}}{er}, \tag{10.36}$$

and

$$u = \cos^{-1} \frac{\vec{n} \cdot \vec{r}}{nr}. \tag{10.37}$$

Finally, the semi-major axis of the orbit can be calculated from the equation of energy (Sec. 2.9):

$$v^2 = GM \left(\frac{2}{r} - \frac{1}{a} \right). \tag{10.38}$$

Solving, one gets

$$a = \frac{1}{\frac{2}{r} - \frac{v^2}{GM}}. \tag{10.39}$$

An alternative measure of a is furnished by Eq. (9.36):

$$p = \frac{l^2}{GMm^2} = \frac{h^2}{GM}, \tag{10.40}$$

from which

$$a = \frac{h^2}{GM(1 - e^2)}. \tag{10.41}$$

Frequently, the number of revolutions of the satellite per day (24 h), N (loosely called the "mean motion") is given, in which case, the period P (in hr) of the satellite is 24/N. The real mean anomaly n in rad/s (not to be confused with the magnitude of the node vector) is given by

$$n = \frac{2\pi}{P} = \frac{2\pi N}{24 \times 60 \times 60}. \tag{10.42}$$

Equation (3.40) of Sec. 3.4 furnishes the semi-major axis

$$a = \sqrt[3]{\frac{GM}{n^2}}. \tag{10.43}$$

10.7 The Topocentric-Horizontal Coordinate System

The position and velocity vectors of the satellite in Sec. 10.6 refer to the inertial geocentric-equatorial coordinate system. In reality, radar measurements are taken from a ground station on the surface

of the Earth, which is a rotating coordinate system, and therefore
not inertial. The measured position and velocity of the satellite have
to be converted into the geocentric inertial coordinate system. In
the ground-based coordinate system (x', y', z') with the radar site as
origin, the vertical is normally chosen as the z'-axis; the southward
direction is the x'-axis; and the eastward direction is the y'-axis
(Fig. 10.3). Such a system is referred to as the ***topocentric-horizontal
coordinate system***. Let the observed position and velocity vectors in
this system be $\vec{r}'$ and $\vec{v}'$ respectively, where

$$\vec{v}' = \frac{d\vec{r}'}{dt}. \tag{10.44}$$

If $\vec{r}$ is the position vector of the satellite and $\vec{R}$ is the position vector
of the radar site in the geocentric-equatorial system, then by vector
addition (vide Fig. 10.3)

$$\vec{r} = \vec{r}' + \vec{R}. \tag{10.45}$$

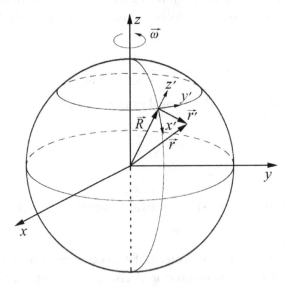

Fig. 10.3

For the "true" velocity of the satellite in the geocentric inertial frame, we have to add the velocity of the point in the topocentric-horizontal system where the satellite is instantaneously located [cf. Bate *et al.* (1971)],

$$\vec{v} = \vec{v}' + \vec{\omega} \times \vec{r},\qquad(10.46)$$

where $\vec{\omega}$ is the angular velocity of rotation of the Earth.

10.8 Approximate Solution of Kepler's Equation

The true anomaly θ of the satellite is traditionally determined from the mean anomaly M via the eccentric anomaly E. The transformation from E to θ is facilitated by Eq. (3.29). We have

$$\theta = 2\tan^{-1}\left[\sqrt{\frac{1+e}{1-e}}\tan\frac{E}{2}\right].\qquad(10.47)$$

However, the conversion of M to E involves the solution of Kepler's equation,

$$E - e\sin E = M.\qquad(10.48)$$

Kepler's equation is a transcendental equation, whose solution engaged the greatest mathematical minds of the past. Many approximate solutions have been advanced, but no exact solutions in closed form have been found. For small eccentricities, series solutions of high accuracy can be obtained in only a few terms.

For $e = 0$, we have, in the zeroeth approximation

$$E = M.\qquad(10.49)$$

Putting $E = M$ in Eq. (10.48), we get the solution in the first approximation [cf. Van de Kamp (1964)],

$$E = M + e\sin M.\qquad(10.50)$$

In the second approximation, we have

$$E = M + e \sin (M + e \sin M). \tag{10.51}$$

Expanding, and retaining terms up to e^2,

$$E = M + e \sin M + \frac{1}{2} e^2 \sin 2M. \tag{10.52}$$

If we keep terms up to e^3, we get the third-order approximation [cf. Danby (1988)],

$$E = M + e \sin M + \frac{1}{2} e^2 \sin 2M + \frac{1}{8} e^3 (3 \sin 3M - \sin M). \tag{10.53}$$

10.9 Ground Track of a Satellite

The **ground track** of a satellite is the locus of the foot of the perpendicular from the satellite traced out on the ground. It is normally plotted on a Cartesian world map with longitude as abscissa and co-latitude as ordinate. Due to the rotation of the Earth, the ground track drifts westwards after each revolution. It crosses the equator twice during one revolution. The greatest latitude attained by the ground track both north and south of the equator is equal to the inclination i of the satellite (for $i < 90°$) or its complement from $180°$, i.e., $\pi - i$ (for $i > 90°$). Thus, regions of the Earth having latitudes greater than the inclination i cannot be covered from LEO satellites. Ground tracks of various types of satellites are elegantly illustrated in Verger *et al.* (2003).

Figure 10.4 is an example of the ground track of a satellite. It belonged to the unseparated Titan rocket/satellite on its final orbit when it fragmented as a result of an explosion on October 15, 1965. At that time, the satellite had an inclination of $32°$ and period of 100 min.

The latitude of the launch site (north or south) sets both a lower limit and an upper limit to the inclination of the satellite

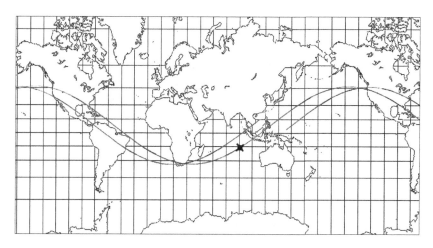

Fig. 10.4

launched. The lower limit of the inclination i achieved is equal to the latitude λ of the launch site, which can be achieved by launching the satellite eastwards. Thus, for satellites launched from Cape Canaveral (latitude 28.4°), the minimum inclination is 28.4°. If the satellite is launched westwards, the upper limit of inclination equal to 180° — λ is attained.

A *Sun-synchronous satellite* is one which passes over a given location on the Earth's surface at the same local time. It must make an even number of revolutions each solar day (24 h). A typical Sun-synchronous satellite, making 15 revolutions per day has an inclination of approximately 98° and a period of 96 min. Such a satellite will pass over 15 longitudinal sectors each day.

Exercises

10.1. Derive Eqs. (10.3) and (10.4).
10.2. Verify Eq. (10.9).
10.3. Verify Eqs. (10.20) to (10.22).
10.4. Verify Eqs. (10.24) to (10.26).

10.5. Show that $e = \frac{h_A - h_P}{2r_0 + h_A + h_P}$.

10.6. Express b in terms of h_A and h_P.

10.7. Express p in terms of h_A and h_P.

10.8. Sometimes radial distances of the apogee and perigee are used instead of the heights of the apsidal points [cf. Nelson and Loft (1962)]:

$$r_A = r_0 + h_A, \quad \text{and}$$
$$r_P = r_0 + h_P.$$

(a) Express a in terms of r_A and r_P.

(b) Express e in terms of r_A and r_P.

(c) Express b in terms of r_A and r_P.

(d) Express p in terms of r_A and r_P.

10.9. Express the number of revolutions of a geosynchronous satellite per day N in terms of the solar day P (24 h) and the sidereal rotation period of the Earth P_s (23.9345 h).

10.10. Verify Eqs. (10.50) to (10.53).

10.11. Estimate the angular drift of the Titan satellite from its period.

10.12. Discuss the ground tracks of geostationary and geosynchronous satellites.

Perturbations of
Satellite Orbits

11.1 Perturbations of Satellite Orbits

If the one-body approximation for satellite motion held true and if there were no external perturbing forces on the satellite whatsoever, then one would expect the orbit to be purely Keplerian. The orbital elements (with the exception of true anomaly) would remain constants. However, departures from the ideal situation are the reality. Satellite orbits are always affected by atmospheric drag, which causes their eventual decay. The oblateness of the Earth produces periodic changes in the eccentricity. Satellite orbits can, of course, be altered by mechanical impulses applied to them. Smaller effects are created by gravitational torques produced by the Moon and the Sun.

The rates of changes of the orbital elements due to *perturbative forces* have been worked out by Lagrange, Encke and Cowell, among others. Their methods are still used today. The methods of Encke and Cowell are numerical in nature and are useful in today's high-speed computational techniques. In the following sections, we shall discuss the analytical approach of Lagrange and the resulting equations known as *Lagrange's planetary equations.*

11.2 Perturbative Forces on a Satellite in Orbit

It is easier to analyze the effects of perturbative forces on the orbital elements of a satellite in the satellite's local frame of reference. A natural coordinate system (**System 1**) is one in which one axis is in the radial direction from the attracting center, the second axis is in the transverse direction in the direction of increasing true anomaly and the third axis is in the direction of the orbital angular momentum of the satellite [cf. Bate *et al.* (1971)]. In such a system, the general expression for the perturbative force is

$$\vec{F} = F_r \hat{r} + F_\theta \hat{\theta} + F_h \hat{h}, \tag{11.1}$$

where F_r is the *radial component* of the force, F_θ the *transverse component* (also called the *down-range component*) and F_h is the *orthogonal component* (also called the *cross-range component*). F_r and F_θ lie in the plane of the orbit, whereas F_h is perpendicular to the orbital plane. The unit vectors $\hat{r}, \hat{\theta}$ and $\hat{h}$ form a right-handed system.

In an alternative coordinate system (**System 2**), one axis is along the velocity vector in the direction tangential to the orbit, the second axis is along the normal direction and the third axis is in the direction of the orbital angular momentum [cf. Ehricke (1962)]. In this system, the perturbative force takes the form

$$\vec{F} = F_s \hat{s} + F_t \hat{t} + F_h \hat{h}, \tag{11.2}$$

where F_s is the *tangential component* of the force, F_t is the *normal component* and F_h is the *orthogonal component*. F_s and F_t are in the plane of the orbit while F_h is perpendicular to the orbital plane. The unit vectors $\hat{t}, \hat{s}$ and $\hat{h}$ form a second right-handed system.

The transformation equations for the force components from the Systems 1 to 2 coordinates are the same as those for the jerk components in Sec. 3.8:

$$F_s = F_r \sin \phi + F_\theta \cos \phi, \tag{11.3}$$

and

$$F_t = F_r \cos \phi - F_\theta \sin \phi, \tag{11.4}$$

with

$$\sin \phi = \frac{e \sin \theta}{\sqrt{1 + e^2 + 2e \cos \theta}}, \tag{11.5}$$

and

$$\cos \phi = \frac{1 + e \cos \theta}{\sqrt{1 + e^2 + 2e \cos \theta}}. \tag{11.6}$$

The inverse transformations for the force components are obtained by elimination between Eqs. (11.3) and (11.4):

$$F_r = F_s \sin \phi + F_t \cos \phi, \tag{11.7}$$

and

$$F_\theta = F_s \cos \phi - F_t \sin \phi. \tag{11.8}$$

It should be noted that neither of these local frames of references is an inertial system. However, it is easier to analyze the effects of the perturbative forces in these systems. In the following sections, we shall discuss the rates of changes of orbital elements due to arbitrary perturbative forces.

11.3 Effect of a Perturbing Force on the Semi-Major Axis

The change in energy of the orbit is equal to the work done by the perturbing force,

$$dE = \vec{F} \cdot d\vec{r}, \tag{11.9}$$

whence

$$\frac{dE}{dt} = \vec{F} \cdot \frac{d\vec{r}}{dt} = \vec{F} \cdot \vec{v}. \tag{11.10}$$

From Eq. (2.87), we have

$$E = -\frac{GMm}{2a}. \tag{11.11}$$

By differentiation using the chain rule,

$$\frac{dE}{dt} = \frac{dE}{da}\frac{da}{dt} = \frac{GMm}{2a^2}\frac{da}{dt}. \tag{11.12}$$

Thus, the rate of change of the semi-major axis due to the perturbing force is

$$\frac{da}{dt} = \frac{2a^2}{GMm}\frac{dE}{dt}. \tag{11.13}$$

By virtue of Eq. (11.10),

$$\frac{da}{dt} = \frac{2a^2}{GMm}\vec{F} \cdot \vec{v}. \tag{11.14}$$

We have, in System 1,

$$\vec{F} \cdot \vec{v} = \frac{dr}{dt}F_r + r\frac{d\theta}{dt}F_\theta = \frac{d\theta}{dt}\left[\frac{dr}{d\theta}F_r + rF_\theta\right]. \tag{11.15}$$

For the factor $d\theta/dt$, we have from the conservation of angular momentum,

$$\frac{d\theta}{dt} = \frac{l}{mr^2}, \tag{11.16}$$

where the angular momentum can be expressed from Eq. (4.18):

$$l^2 = GMm^2p, \tag{11.17}$$

with

$$p = a(1 - e^2) \tag{11.18}$$

Substituting Eqs. (11.17) and (11.18) in Eq. (11.16), we get

$$\frac{d\theta}{dt} = \frac{\sqrt{GMa}}{r^2}\sqrt{1 - e^2}. \tag{11.19}$$

For the factor $dr/d\theta$, we utilize the equation of the orbital ellipse

$$r = \frac{p}{1 + e\cos\theta}. \tag{11.20}$$

By differentiation,

$$\frac{dr}{d\theta} = \frac{pe\sin\theta}{(1 + e\cos\theta)^2} = \frac{r^2 e\sin\theta}{a(1 - e^2)}. \tag{11.21}$$

It is customary to convert GM into the mean motion n of the satellite via Eq. (3.40):

$$n = \sqrt{\frac{GM}{a^3}}. \tag{11.22}$$

Thus, Eqs. (11.14) and (11.19) become

$$\frac{da}{dt} = \frac{2}{mn^2a}\vec{F}\cdot\vec{v}, \tag{11.23}$$

and

$$\frac{d\theta}{dt} = \frac{na^2}{r^2}\sqrt{1 - e^2}. \tag{11.24}$$

By combining Eqs. (11.15), (11.21), (11.23) and (11.24), we finally arrive at the standard expression for the rate of change of the semi-major axis due to a perturbative force in System 1 of the satellite's own frame of reference [cf. Bate *et al.* (1971)],

$$\frac{da}{dt} = \frac{2}{mn}\left[\frac{e\sin\theta}{\sqrt{1-e^2}}F_r + \frac{a\sqrt{1-e^2}}{r}F_\theta\right].$$ (11.25)

One can obtain this result in System 2 coordinates by converting the force components by means of Eqs. (11.7), (11.8), (11.5) and (11.6), whence Eq. (11.25) transforms to

$$\frac{da}{dt} = \frac{2}{mn}\frac{\sqrt{1+e^2+2e\cos\theta}}{\sqrt{1-e^2}}F_s.$$ (11.26)

By virtue of Eqs. (3.71) and (11.17), this equation reduces to [cf. Ehricke (1962)]

$$\frac{da}{dt} = \frac{2}{mn^2a}vF_s.$$ (11.27)

Since the velocity of the satellite in entirely in the tangential direction,

$$\vec{v} = v\hat{s}.$$ (11.28)

Hence,

$$\vec{v}\cdot\vec{F} = vF_s.$$ (11.29)

Equation (11.27) thus tells us that the semi-major axis can only be altered by the tangential component of the perturbing force and not by the normal component or the orthogonal component. A perturbing force in the forward direction increases the semi-major axis, and therefore, the energy of the orbit, whereas a perturbing force in the backward direction decreases the semi-major axis and the energy of the orbit. Furthermore, since the speed of the satellite is the greatest at its perigee and the smallest at the apogee, the greatest gain in energy takes place when the forward perturbing force acts at the perigee, whereas the least gain is achieved at the apogee. These results are of paramount importance in the execution of orbital transfers.

Alternatively, one can obtain Eq. (11.27) directly from Eqs. (11.14), (11.22) and (11.29). Likewise, Eq. (11.15) can be obtained from Eq. (11.27) by means of Eqs. (3.71), (11.17) and (11.22).

11.4 Effect of a Perturbing Force on Eccentricity

We begin with Eqs. (11.17) and (11.18),

$$e = \sqrt{1 - \frac{p}{a}} = \sqrt{1 - \frac{l^2}{GMm^2a}}. \tag{11.30}$$

Differentiating with respect to time, we have

$$\frac{de}{dt} = -\frac{l}{GMm^2ae}\frac{dl}{dt} + \frac{l^2}{2GMm^2a^2e}\frac{da}{dt}. \tag{11.31}$$

Converting l and GM into the mean motion n by means of Eqs. (11.17), (11.18) and (11.22),

$$\frac{de}{dt} = -\frac{\sqrt{1 - e^2}}{mna^2e}\frac{dl}{dt} + \frac{1 - e^2}{2ae}\frac{da}{dt}. \tag{11.32}$$

Now, the rate of change of the angular momentum due to the perturbative force is given by

$$\frac{d\vec{l}}{dt} = \vec{r} \times \vec{F} = \begin{vmatrix} \hat{r} & \hat{\theta} & \hat{h} \\ r & 0 & 0 \\ F_r & F_\theta & F_h \end{vmatrix} = rF_\theta\hat{h} - rF_h\hat{\theta}. \tag{11.33}$$

Also,

$$\frac{d\vec{l}}{dt} = \frac{dl}{dt}\hat{h} + l\frac{d\hat{h}}{dt}. \tag{11.34}$$

Comparing Eqs. (11.33) and (11.34), we get

$$\frac{dl}{dt} = rF_\theta. \tag{11.35}$$

Equation (11.32) gives, with substitutions from Eqs. (11.25) and (11.35), the rate of change of eccentricity due to the perturbative force in System 1 [cf. Bate *et al.* (1971)]:

$$\frac{de}{dt} = \frac{\sqrt{1-e^2}}{mna}\left\{\sin\theta F_r + \left[\frac{a(1-e^2)}{er} - \frac{r}{ae}\right]F_\theta\right\}. \quad (11.36)$$

Converting the force components from Eqs. (11.5)–(11.8) and reducing quantities by Eqs. (3.71) and (11.17), we obtain the rate of change of eccentricity in System 2 [cf. Ehricke (1962)]:

$$\frac{de}{dt} = \frac{2(e+\cos\theta)}{mv}F_s + \frac{r\sin\theta}{mav}F_t. \quad (11.37)$$

Equations (11.36) and (11.37) indicate that at the ends of the major axis ($\theta = 0$ and $\theta = \pi$), F_r and F_t have no effect in altering the eccentricity of the orbit. Also, at the ends of the minor axis, $\theta = \cos^{-1} e$, or $e + \cos\theta = 0$, which means that F_s has no effect in altering the eccentricity there. Furthermore, $\cos\theta$ is maximum at the ends of the latus rectum through the attracting center. Hence F_s has the maximum effect of increasing the eccentricity at these points in accordance with Eq. (11.37).

11.5 Effect of a Perturbing Force on Inclination

From Eq. (10.30), we had

$$\cos i = \frac{\vec{h}\cdot\hat{z}}{h} = \frac{\vec{l}\cdot\hat{z}}{l}. \quad (11.38)$$

Differentiating both sides with respect to time, we get

$$-\sin i\frac{di}{dt} = \frac{l\frac{d\vec{l}}{dt}\cdot\hat{z} - \vec{l}\cdot\hat{z}\frac{dl}{dt}}{l^2}. \quad (11.39)$$

Substituting from Eqs. (11.33) and (11.35), we get

$$-\sin i\frac{di}{dt} = -\frac{rF_h\hat{\theta}\cdot\hat{z}}{l}. \quad (11.40)$$

Now (vide Fig. 10.2),

$$\hat{\theta} \cdot \hat{z} = \sin i \cos u. \tag{11.41}$$

Hence,

$$\frac{di}{dt} = \frac{rF_h \cos u}{l}. \tag{11.42}$$

By converting l into the orbital elements by means of Eqs. (11.17), (11.18) and (11.22), we arrive at the rate of change of inclination in either system [cf. Bate *et al.* (1971)]:

$$\frac{di}{dt} = \frac{r \cos u}{mna^2\sqrt{1 - e^2}} F_h. \tag{11.43}$$

Thus, the inclination of the satellite is affected only by the orthogonal component of the perturbing force and not by any other component in the plane of the orbit. The rate of change of inclination is proportional to the cosine of the argument of the perigee, which is the angular position of the satellite measured from the ascending node (Fig. 10.2). For maximum change of inclination, the orthogonal component of force has to be applied at the nodal points ($u = 0$ and $u = \pi$).

11.6 Effect of a Perturbing Force on the Longitude of an Ascending Node

The longitude of the ascending node Ω, like the inclination i, can only be altered by the orthogonal component of the perturbing force F_h and not by any other component in the plane of the orbit. We begin with Eqs. (10.10) and (10.31) [cf. Bate *et al.* 1871]:

$$\cos \Omega = \frac{\hat{x} \cdot \vec{n}}{|\vec{n}|} = \frac{\hat{x} \cdot (\hat{z} \times \vec{h})}{|\hat{z} \times \vec{h}|} = \frac{\hat{x} \cdot (\hat{z} \times \vec{l})}{|\hat{z} \times \vec{l}|}. \tag{11.44}$$

Differentiating both sides with respect to time, we get

$$-\sin\Omega\frac{d\Omega}{dt} = \frac{\hat{x}\cdot\left(\hat{z}\times\frac{d\vec{l}}{dt}\right)|\hat{z}\times\vec{l}| - \hat{x}\cdot(\hat{z}\times\vec{l})\frac{d}{dt}|\hat{z}\times\vec{l}|}{|\hat{z}\times\vec{l}|^2}. \tag{11.45}$$

Upon expanding and substituting from Eq. (11.33), we get

$$-\sin\Omega\frac{d\Omega}{dt} = \frac{A-B}{l^2\sin^2 i}, \tag{11.46}$$

where

$$A = \hat{x}\cdot[\hat{z}\times(rF_\theta\hat{h} - rF_h\hat{\theta})]\,l\sin i, \tag{11.46.1}$$

and

$$B = l\cos\Omega\sin i\left(\frac{dl}{dt}\sin i + l\cos i\frac{di}{dt}\right). \tag{11.46.2}$$

In Eq. (11.46), we have already used the result

$$\hat{x}\cdot(\hat{z}\times\vec{l}) = l\cos\Omega\sin i. \tag{11.47}$$

Furthermore,

$$\hat{x}\cdot(\hat{z}\times\hat{\theta}) = (\hat{x}\times\hat{z})\cdot\hat{\theta} = -\hat{y}\cdot\hat{\theta}. \tag{11.48}$$

The last quantity can be determined from the Euler angle transformation of the axes. Notice that the $(\hat{r}, \hat{\theta}, \hat{h})$ system is obtained through three successive rotations of the (x, y, z) coordinates: firstly, a counter-clockwise rotation about the z-axis through angle Ω; secondly, a rotation about the nodal line through an angle i; and lastly, a rotation about the angular momentum vector by an angle u (Fig. 11.1). From the Euler angle transformation, we get [cf. Nelson and Loft (1962)]

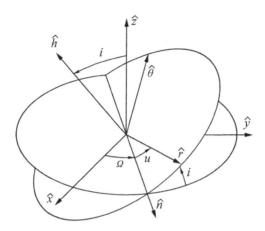

Fig. 11.1

$$\begin{pmatrix} \hat{r} \\ \hat{\theta} \\ \hat{h} \end{pmatrix} = \begin{pmatrix} a_{11} & a_{12} & a_{13} \\ a_{21} & a_{22} & a_{23} \\ a_{31} & a_{32} & a_{33} \end{pmatrix} \begin{pmatrix} \hat{x} \\ \hat{y} \\ \hat{z} \end{pmatrix}. \tag{11.49}$$

Here the components of the rotation matrix are:

$$a_{11} = \cos u \cos i - \cos i \sin \Omega \sin u, \tag{11.49.1}$$

$$a_{12} = \cos u \sin \Omega + \cos i \cos \Omega \sin u, \tag{11.49.2}$$

$$a_{13} = \sin u \sin i, \tag{11.49.3}$$

$$a_{21} = -\sin u \cos \Omega - \cos i \sin \Omega \cos u, \tag{11.49.4}$$

$$a_{22} = -\sin u \sin \Omega + \cos i \cos \Omega \cos u, \tag{11.49.5}$$

$$a_{23} = \cos u \sin i, \tag{11.49.6}$$

$$a_{31} = \sin i \sin \Omega, \tag{11.49.7}$$

$$a_{32} = -\sin i \cos \Omega, \tag{11.49.8}$$

and

$$a_{33} = \cos i. \tag{11.49.9}$$

Taking the dot-product of the middle equation of (11.49) with $\hat{y}$, we get

$$\hat{\theta} \cdot \hat{y} = -\sin u \sin \Omega + \cos i \cos \Omega \cos u. \tag{11.50}$$

Substituting Eqs. (11.35), (11.43) and (11.50) into Eq. (11.46), simplifying and converting into the orbital elements by Eqs. (4.18) and (11.15), we finally arrive at [cf. Bate *et al.* (1971)]

$$\frac{d\Omega}{dt} = \frac{r \sin u}{mna^2\sqrt{1 - e^2} \sin i} F_h.$$ (11.51)

The rate of change of Ω is maximized when $u = \pi/2$ or $3\pi/2$, i.e., when the satellite is farthest from the nodal line. Also, due to the presence of $\sin i$ in the denominator, the change is amplified for low-inclination orbits.

11.7 Effect of a Perturbing Force on True Anomaly

We commence with the important statement that since $\vec{r}$ is not instantaneously altered by the perturbing force, its first derivative is zero [cf. Bate *et al.* (1971); Danby (1988)]. First, from Eq. (4.18), the equation of the orbital ellipse is written as

$$r(1 + e\cos\theta) = \frac{l^2}{GMm^2}.$$ (11.52)

Differentiating with respect to time,

$$r\cos\theta\frac{de}{dt} - re\sin\theta\frac{d\theta}{dt} = \frac{2l}{GMm^2}\frac{dl}{dt}.$$ (11.53)

Next, taking the dot-product of Eq. (1.63) with $\vec{r}$, we have

$$GMmre\sin\theta = l\vec{r} \cdot \vec{v}.$$ (11.54)

By differentiation with respect to time,

$$GMmr\sin\theta\frac{de}{dt} + GMmre\cos\theta\frac{d\theta}{dt} = \vec{r} \cdot \vec{v}\frac{dl}{dt} + l\vec{r} \cdot \frac{d\vec{v}}{dt}.$$ (11.55)

Eliminating de/dt between Eqs. (11.53) and (11.55), we get

$$GMmre\frac{d\theta}{dt} = l\cos\theta\vec{r} \cdot \frac{d\vec{v}}{dt} - \frac{2l\sin\theta}{m}\frac{dl}{dt} + \cos\theta\vec{r} \cdot \vec{v}\frac{dl}{dt}.$$ (11.56)

From Eq. (11.1),

$$\vec{r} \cdot \frac{d\vec{v}}{dt} = \frac{r}{m} F_r. \tag{11.57}$$

Also, by Eq. (11.35),

$$\frac{dl}{dt} = r F_\theta. \tag{11.58}$$

Substituting Eqs. (11.57) and (11.58) into (11.56), simplifying, and converting the quantities into the orbital elements as usual, one arrives at [cf. Bate *et al.* (1971)]

$$\frac{d\theta}{dt} = \frac{\sqrt{1 - e^2}}{mnae} \left[\cos\theta F_r - \sin\theta \left(1 + \frac{1}{1 + e\cos\theta} \right) F_\theta \right]. \tag{11.59}$$

We can obtain the equation in System 2 by converting F_r and F_θ with the help of Eqs. (11.5)–(11.8). Using Eqs. (3.71) and (11.17), we get [cf. Ehricke (1962)]

$$\frac{d\theta}{dt} = -\frac{2\sin\theta}{mve} F_s + \frac{2ae + r\cos\theta}{maev} F_t. \tag{11.60}$$

11.8 Effect of a Perturbing Force on the Argument of Perigee

The argument of perigee ω is intimately connected to the true anomaly θ via the argument of latitude u. As such, its rate of change due to a perturbing force is related to that of the true anomaly. We begin with Eq. (10.34):

$$r\cos(\omega + \theta) = \frac{\vec{n} \cdot \vec{r}}{n} = \frac{(\hat{z} \times \vec{l}) \cdot \vec{r}}{\left| \hat{z} \times \vec{l} \right|}. \tag{11.61}$$

We recall, as before, that $\vec{r}$ does not change to the first-order due to the perturbing force. Differentiating Eq. (11.61) with respect to time, we get

$$-r\sin(\omega+\theta)\left(\frac{d\omega}{dt}+\frac{d\theta}{dt}\right)=\frac{\left|\hat{z}\times\vec{l}\right|\left(\hat{z}\times\frac{d\vec{l}}{dt}\cdot\vec{r}\right)-\left(\hat{z}\times\vec{l}\cdot\vec{r}\right)\frac{d}{dt}\left|\hat{z}\times\vec{l}\right|}{\left|\hat{z}\times\vec{l}\right|^{2}}.$$

(11.62)

Substituting from Eq. (11.33) and rearranging, we get

$$\frac{d\omega}{dt}=\frac{-|\hat{z}\times\vec{l}|[\hat{z}\times(rF_{\theta}\hat{h}-rF_{h}\hat{\theta})\cdot\vec{r}]+(\hat{z}\times\vec{l}\cdot\vec{r})\frac{d}{dt}|\hat{z}\times\vec{l}|}{|\hat{z}\times\vec{l}|^{2}\,r\sin(\omega+\theta)}-\frac{d\theta}{dt}.$$

(11.63)

Now, one can verify the following (vide Fig. 10.2):

$$|\hat{z}\times\vec{l}|=l\sin i,\qquad\qquad(11.64)$$

$$\hat{z}\times\vec{l}\cdot\vec{r}=rl\sin i\cos u,\qquad(11.65)$$

$$\hat{z}\times\hat{h}\cdot\vec{r}=r\sin i\cos u,\qquad(11.66)$$

and

$$\hat{z}\times\hat{\theta}\cdot\vec{r}=r\hat{z}\times\hat{\theta}\cdot\hat{r}=r\hat{\theta}\times\hat{r}\cdot\hat{z}=-r\hat{h}\cdot\hat{z}=-r\cos i.\quad(11.67)$$

Also

$$\frac{d}{dt}|\hat{z}\times\vec{l}|=\frac{d}{dt}(l\sin i)=\frac{dl}{dt}\sin i+l\cos i\frac{di}{dt}.\quad(11.68)$$

Substituting Eqs. (11.64)–(11.68) in Eq. (11.63) and putting in values for dl/dt and di/dt from Eqs. (11.35) and (11.42) respectively, we arrive at [cf. Bate *et al.* (1971)]

$$\frac{d\omega}{dt}=-\frac{r\cot i\sin u}{mna^{2}\sqrt{1-e^{2}}}F_{h}-\frac{d\theta}{dt},\quad(11.69)$$

where $d\theta/dt$ is given by Eq. (11.59). The argument of perigee is the only orbital element which is affected by both the in-plane and orthogonal perturbative forces.

11.9 Effects of a Pertubing Force on the Period and Mean Motion

The period of a satellite is related to the semi-major axis via Kepler's third law. From Eq. (1.1), we can write

$$P = k_1 a^{3/2}, \tag{11.70}$$

where k_1 is the constant of proportionality. Differentiating both sides with respect to time,

$$\frac{dP}{dt} = \frac{3}{2} k_1 a^{1/2} \frac{da}{dt}. \tag{11.71}$$

Equations (11.70) and (11.71) give the rate of change of the period

$$\frac{dP}{dt} = \frac{3}{2} \frac{P}{a} \frac{da}{dt}, \tag{11.72}$$

where da/dt is given by Eqs. (11.25) and (11.27).

The mean motion expressed as a function of the semi-major axis is given by Eq. (3.40):

$$n = k_2 a^{-3/2}, \tag{11.73}$$

where k_2 is another constant of proportionality. Differentiating both sides with respect to time, we have

$$\frac{dn}{dt} = -\frac{3}{2} k_2 a^{-5/2} \frac{da}{dt}. \tag{11.74}$$

Once again, from Eqs. (11.73) and (11.74), we get

$$\frac{dn}{dt} = -\frac{3}{2} \frac{n}{a} \frac{da}{dt}. \tag{11.75}$$

11.10 Effects of Finite Impulse on the Orbital Elements

If the perturbing force acts for a short interval dt, thus creating velocity perturbations in the three orthogonal components dv_r, dv_θ and dv_h, then by the impulse-momentum theorem, we can write

$$F_r dt = m dv_r, \qquad (11.76)$$
$$F_\theta dt = m dv_\theta, \qquad (11.77)$$

and

$$F_h dt = m dv_h. \qquad (11.78)$$

Lagrange's planetary equations can then be re-written as [cf. Meirovitch (1970)]

$$da = \frac{2}{n}\left[\frac{e\sin\theta}{\sqrt{1-e^2}}dv_r + \frac{a\sqrt{1-e^2}}{r}dv_\theta\right], \qquad (11.79)$$

$$de = \frac{\sqrt{1-e^2}}{na}\left\{\sin\theta dv_r + \left[\frac{a(1-e^2)}{er} - \frac{r}{ae}\right]dv_\theta\right\}, \qquad (11.80)$$

$$di = \frac{r\cos u}{na^2\sqrt{1-e^2}}dv_h, \qquad (11.81)$$

$$d\Omega = \frac{r\sin u}{na^2\sqrt{1-e^2}\sin i}dv_h, \qquad (11.82)$$

$$d\theta = \frac{\sqrt{1-e^2}}{nae}\left[\cos\theta dv_r - \sin\theta\left(1 + \frac{1}{1+e\cos\theta}\right)dv_\theta\right], \qquad (11.83)$$

and

$$d\omega = -\frac{r\cot i\sin u}{na^2\sqrt{1-e^2}}dv_h - d\theta. \qquad (11.84)$$

If the tangential and normal components of the impulse are considered, then

$$F_s dt = m dv_s, \tag{11.85}$$

and

$$F_t dt = m dv_t. \tag{11.86}$$

In that case,

$$da = \frac{2v}{n^2 a} dv_s, \tag{11.87}$$

$$de = \frac{2(e + \cos\theta)}{v} dv_s - \frac{r\sin\theta}{av} dv_t, \tag{11.88}$$

and

$$d\theta = -\frac{2\sin\theta}{ev} dv_s + \frac{2ae + r\cos\theta}{aev} dv_t. \tag{11.89}$$

11.11 Effects of Atmospheric Drag on Orbital Elements

Like all **drag forces, air resistance** is always anti-parallel to the velocity of an object moving through a fluid medium and is most commonly proportional to the square of the velocity. For atmospheric drag on orbiting satellites, the magnitude of the drag force is customarily written as [cf. King-Hele (1987)]

$$F_D = \frac{1}{2} C_D A \rho v^2, \tag{11.90}$$

where A is the **cross-sectional area** of the satellite at its instantaneous orientation and ρ is the density of air encountered by the satellite. The **drag coefficient** C_D depends on the shape and structure of the leading surface, its roughness, and other properties and has a value of greater than two [cf. King-Hele (1987)]. For satellites of unknown drag coefficient, the value of C_D is generally taken as 2.2 [cf. King-Hele (1987)]. The effects of atmospheric drag on the orbital elements of a satellite can be conveniently analyzed in the System 2 coordinates, where the drag force is simply

$$\vec{F} = -F_D \hat{s}. \tag{11.91}$$

The effect of air drag on the semi-major axis of a satellite is obtained from Eqs. (11.27) and (11.90):

$$\frac{da}{dt} = -\frac{C_D A \rho v^3}{m n^2 a}. \tag{11.92}$$

It is usually more convenient to use the eccentric anomaly E to calculate drag effects. From Eqs. (2.76) and (3.23),

$$v^2 = \frac{GM}{a} \frac{1 + e \cos E}{1 - e \cos E}. \tag{11.93}$$

Furthermore, from Eq. (3.40), we have

$$dt = \frac{a^{3/2}}{\sqrt{GM}} (1 - e \cos E) dE. \tag{11.94}$$

Equation (10.92) gives, with substitutions from Eqs. (10.93) and (10.94),

$$da = -\frac{C_D A a^2 \rho}{m} \frac{(1 + e \cos E)^{3/2} dE}{\sqrt{1 - e \cos E}}. \tag{11.95}$$

The change of the semi-major axis due to the air drag during one orbit is calculated from [cf. Danby (1988)]

$$\Delta a = -\frac{C_D A a^2}{m} \int_0^{2\pi} \rho \frac{(1 + e \cos E)^{3/2}}{\sqrt{1 - e \cos E}} dE. \tag{11.96}$$

Likewise, we can calculate the change of eccentricity due to the atmospheric drag. From Eqs. (11.37) and (11.90),

$$\frac{de}{dt} = -\frac{C_D A \rho (e + \cos \theta) v}{m}. \tag{11.97}$$

From Eqs. (3.21) and (3.23), we find [cf. King-Hele (1987)]

$$e + \cos\theta = \frac{1 - e^2}{1 - e\cos E}\cos E. \qquad (11.98)$$

Substitutions from Eqs. (11.93), (11.94) and (11.98) in Eq. (11.97) gives

$$de = -\frac{C_D A\rho(1 - e^2)}{m}\sqrt{\frac{1 + e\cos E}{1 - e\cos E}}\cos E\,dE. \qquad (11.99)$$

The change of eccentricity is obtained by integrating Eq. (11.99) over one orbit [cf. Danby (1988)]:

$$\Delta e = -\frac{C_D A(1 - e^2)}{m}\int_0^{2\pi}\rho\sqrt{\frac{1 + e\cos E}{1 - e\cos E}}\cos E\,dE. \qquad (11.100)$$

The only other orbital element which is possibly affected by atmospheric drag is the argument of perigee ω. From Eqs. (11.60), (11.69) and (11.90), we find

$$\frac{d\omega}{dt} = -\frac{C_D A\rho v \sin\theta}{me}. \qquad (11.101)$$

However, this effect is small. The integral of $d\omega/dt$ over one orbit is nearly zero [cf. Danby (1988)].

Equations (11.96) and (11.100) indicate that atmospheric drag continually reduces both the semi-major axis and eccentricity of an orbiting satellite. The orbit becomes smaller and rounder after each revolution. It is also interesting to analyze the changes in the apogee and perigee distances from the center of the Earth, which are respectively

$$r_A = a(1 + e), \qquad (11.102)$$

and

$$r_P = a(1 - e). \qquad (11.103)$$

We have

$$\Delta r_A = \Delta a + a\Delta e + e\Delta a, \qquad (11.104)$$

and

$$\Delta r_P = \Delta a - a\Delta e - e\Delta a. \qquad (11.105)$$

Clearly, the apogee distance contracts faster than a and e, whereas the perigee distance contracts at a slower rate than those of a and e. This is especially true for satellite orbits having high eccentricities, since atmospheric density decreases rapidly with altitude and most of the drag takes place near the perigee of the satellite. In that case, the effect of drag is similar to that of a backward impulse applied at the perigee in a Hohmann transfer to a smaller orbit.

We should point out that besides drag, there is *aerodynamic lift* encountered by an orbiting satellite. However, most satellites are not attitude-controlled, which means that lift forces continually change in direction and therefore tend to cancel out [cf. King-Hele (1987)].

Historically, the launch of artificial Earth satellites and their orbital decay provided the earliest information regarding the densities, temperature and other parameters of the upper atmosphere. The standard models of the neutral atmosphere, including the various versions of Jacchia (1971 and 1977), CIRA (COSPAR International Reference Atmosphere, 1972) and the later versions such as MSIS (Hedin, 1983) are all based on the orbital decay of the thousands of satellites launched since 1957.

11.12 Effects of Earth's Gravitational Field on Orbital Elements

If the Earth were perfectly spherically symmetric, its gravitational field at an exterior point would be that of a point mass situated at its center. However, the Earth is spherical only to a first approximation; it is ellipsoidal in the second approximation; and "pear-shaped" in the third approximation. The gravitational potential of the Earth can be expressed as an infinite series of terms in *spherical harmonics*. The Earth's "*oblateness*" is represented by the J_2 term in the harmonic expansion, whereas the *"pear shape"* is given by the J_3 term.

The Earth's oblateness causes major changes in two of the orbital elements of a satellite. First, the longitude of the ascending node Ω rotates at a rate of

$$\frac{d\Omega}{dt} = -9.97 \left(\frac{R}{a}\right)^{7/2} \frac{\cos i}{(1 - e^2)^2} \qquad (11.106)$$

degrees per day (King-Hele, 1987). In Eq. (11.106), R is the equatorial radius of the Earth. As a result, the orbital plane rotates about the Earth's axis in the direction opposite to the satellite's motion (for $i < 90°$). Second, the argument of perigee ω changes at the rate of

$$\frac{d\omega}{dt} = 4.98 \left(\frac{R}{a}\right)^{7/2} \frac{5\cos^2 i - 1}{(1 - e^2)^2} \qquad (11.107)$$

degrees per day (King-Hele, 1987). This means that the major axis rotates in the plane of the orbit. The advance of perigee is forwards for orbits having an inclination of $i < \cos^{-1}(5^{-1/2}) = 63.435°$ and retrograde for $i > 63.435°$.

Telecommunication satellites of the ex-Soviet Union and Russia use the **Molniya orbit**, which have apogee heights above the geosynchronous orbit; perigee heights above the LEO orbit; periods equal to nearly half of the sidereal roation period of the Earth (12 h); and inclinations around $63.435°$ [cf. Verger *et al.* (2003)]. Their high eccentricities ($e = 0.75$) ensure that the satellites spend most of their time in the northern hemisphere to provide coverage of high latitude areas, while their inclinations ensure that their apogees do not drift in time.

The "pear shape" effect causes periodic oscillations in the perigee distance given by (King-Hele, 1964)

$$r_P = r_{PE} - 6.8 \sin i \sin \omega, \qquad (11.108)$$

where r_{PE} is the equatorial value of r_P and the two distances are measured in km (King-Hele, 1964). Thus r_P has the least value at the most northerly point ($\omega = 90°$) and is the greatest at the most southerly point ($\omega = 270°$).

The oscillation of r_P causes periodic changes in the eccentricity. Since

$$e = 1 - \frac{r_P}{a}, \tag{11.109}$$

the value of the eccentricity when the perigee is at the equator ($\omega = 0°$ or $180°$) is given by

$$e_E = 1 - \frac{r_{PE}}{a}. \tag{11.110}$$

Thus, from Eq. (11.108), one obtains (Tan, 1995)

$$e = e_E + \frac{6.8}{a} \sin i \sin \omega, \tag{11.111}$$

where a is expressed in km.

11.13 Satellite Fragmentation and Orbital Debris

Between 1961 and 2003, a total of 173 satellites have fragmented in orbit (Johnson *et al.* 2004). Most of the fragmentations were explosions of rocket bodies due to ignition of residual fuel by solar radiation; many were due to deliberate detonations by the former Soviet Union (Johnson, 1983a); at least one was the result of a U.S. Anti-satellite (ASAT) experiment [cf. Tan *et al.* (1996)]; two were a part of the U.S. Defense Department's Delta-180 collision experiment [cf. Tan and Zhang (2001)]; and a few were suspected to have been associated with the Soviet ASAT programs of the past (Johnson, 1983b). Yet, the causes for a substantial fraction of the breakups are still unknown (Johnson *et al.* 2004). A necessary by-product of satellite fragmentations is the creation of a large number of hazardous space debris which poses an ever-increasing threat of hyper-velocity collisions to the functioning satellites and space missions. Two natural collisions involving orbital debris have already been reported (Johnson, 1996; Orbital Debris Quarterly News, 2002). Evidence of a major collision has also been advanced (Tan and Ramachandran, 2005).

Discerning the cause of a satellite fragmentation in space has been and continues to be a matter of utmost importance in space debris studies. Pioneering efforts in this direction were undertaken by Culp and McKnight [cf. Johnson and McKnight (1987)] and furthered by McKnight [cf. Johnson and McKnight (1987)]. Based on ground-based experiments involving explosions and hypervelocity collisions, the authors developed methods to distinguish between three classes of satellite fragmentations: (1) low-intensity explosions resulting from combustion; (2) high-intensity explosions resulting from detonations; and (3) hypervelocity collisions. Of major importance was the analysis of the velocity perturbations received by the fragments in the breakup. The latter attained a breakthrough when exact solutions for the velocity perturbations were obtained by R. C. Reynolds [cf. Badhwar *et al.* (1990)].

11.14 Velocity Perturbations from Orbital Element Changes

The magnitude, direction and distribution of the velocity perturbations of the fragments can furnish valuable information regarding the nature and intensity of the fragmentation. The velocity perturbations can be determined from the changes in the orbital elements of the fragments. The procedure is the inverse of that of Sec. 11.10, where the changes in the orbital elements were calculated from the velocity perturbations imparted by finite impulses. This is most conveniently done in the parent satellite's local frame of reference (System 1) at the point of breakup [cf. Badhwar *et al.* (1990)]. Determining the three orthogonal components of the velocity change requires at least three independent equations containing three known quantities, which are normally the changes in the semi-major axis, eccentricity and inclination of the fragment from its parent.

In the System 1 coordinates, the velocity vector of the parent satellite is written as

$$\vec{v} = v_r \hat{r} + v_\theta \hat{\theta}. \tag{11.112}$$

Upon fragmentation, the velocity of a fragment is given by

$$\vec{v}' = (v_r + dv_r)\hat{r} + (v_\theta + dv_\theta)\hat{\theta} + dv_h \hat{h}, \tag{11.113}$$

where dv_r, dv_θ and dv_h are the velocity perturbations of the fragment to be determined (Fig. 11.2). The speeds of the parent and the fragment are, respectively,

$$v = \sqrt{v_r^2 + v_\theta^2}, \tag{11.114}$$

and

$$v' = \sqrt{(v_r + dv_r)^2 + (v_\theta + dv_\theta)^2 + dv_h^2}. \tag{11.115}$$

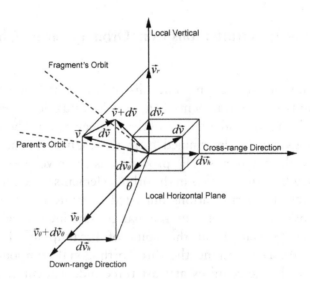

Fig. 11.2

The specific angular momentum of the parent satellite is

$$\vec{h} = \begin{vmatrix} \hat{r} & \hat{\theta} & \hat{h} \\ r & 0 & 0 \\ v_r & v_\theta & 0 \end{vmatrix} = r v_\theta \hat{h}, \tag{11.116}$$

and that of the fragment is

$$\vec{h}' = \begin{vmatrix} \hat{r} & \hat{\theta} & \hat{h} \\ r & 0 & 0 \\ v_r + dv_r & v_\theta + dv_\theta & dv_h \end{vmatrix} = -r dv_h \hat{\theta} + r(v_\theta + dv_\theta)\hat{h}, \tag{11.117}$$

whence

$$h^2 = r^2 v_\theta^2, \tag{11.118}$$

and

$$h'^2 = r^2[(v_\theta + dv_\theta)^2 + dv_h^2]. \tag{11.119}$$

Furthermore, from Eqs. (11.17) and (11.18), we get

$$h^2 = GMa(1 - e^2), \tag{11.120}$$

and

$$h'^2 = GMa'(1 - e'^2). \tag{11.121}$$

From the energy equation [Eq. (2.76)] of the fragment, we have

$$v'^2 = GM\left(\frac{2}{r} - \frac{1}{a'}\right). \tag{11.122}$$

Substituting from Eqs. (11.115), (11.119) and (11.121) and taking the positive root (the negative root corresponds to an extraneous solution), we obtain the perturbation of the radial component of the velocity (Badhwar *et al.*, 1990),

$$dv_r = \sqrt{GM \left(\frac{2}{r} - \frac{1}{a'} \right) - \frac{GM}{r^2} a'(1 - e'^2)} - v_r. \qquad (11.123)$$

Next, the **plane-change angle** ς is defined as the angle between the perturbed and unperturbed orbits. From Fig. 11.2,

$$\tan \varsigma = \frac{dv_h}{v_\theta + dv_\theta}. \qquad (11.124)$$

Then

$$\cos \varsigma = \frac{v_\theta + dv_\theta}{\sqrt{(v_\theta + dv_\theta)^2 + dv_h^2}}, \qquad (11.125)$$

and

$$\sin \varsigma = \frac{dv_h}{\sqrt{(v_\theta + dv_\theta)^2 + dv_h^2}}. \qquad (11.126)$$

By substitution of Eqs. (11.119) and (11.121) into Eqs. (11.125) and (11.126), we arrive at the remaining components of the velocity perturbation of the fragment (Badhwar *et al.*, 1990):

$$dv_\theta = \frac{\cos \varsigma}{r} \sqrt{GMa'(1 - e'^2)} - v_\theta, \qquad (11.127)$$

and

$$dv_h = \frac{\sin \varsigma}{r} \sqrt{GMa'(1 - e'^2)}. \qquad (11.128)$$

In Eqs. (11.127) and (11.128), the plane-change angle ς is as yet undetermined. It can be expressed as a function of the inclination of the parent i, that of the fragment i' and the latitude of the breakup point λ as follows. In Fig. 11.3, we apply the cosines laws to the

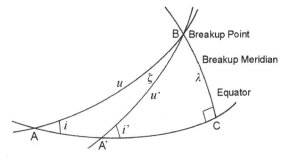

Fig. 11.3

spherical triangle $AA'B$:

$$\cos \varsigma = -\cos i \cos (\pi - i') + \sin i \sin (\pi - i') \cos \phi, \qquad (11.129)$$

and

$$\cos \phi = \cos u \cos u' + \sin u \sin u' \cos \varsigma. \qquad (11.130)$$

Eliminating ϕ between Eqs. (11.129) and (11.130)

$$\cos \varsigma = \pm \frac{\cos i \cos i' + \sin i \sin i' \cos u \cos u'}{1 - \sin i \sin i' \sin u \sin u'}. \qquad (11.131)$$

Applying sine laws to the spherical triangles ABC and $A'BC$, we get

$$\sin u = \frac{\sin \lambda}{\sin i}, \qquad (11.132)$$

and

$$\sin u' = \frac{\sin \lambda}{\sin i'}, \qquad (11.133)$$

whence

$$\cos u = \sqrt{1 - \frac{\sin^2 \lambda}{\sin^2 i}}, \qquad (11.134)$$

and

$$\cos u' = \sqrt{1 - \frac{\sin^2 \lambda}{\sin^2 i'}}. \qquad (11.135)$$

Putting Eqs. (11.134) and (11.135) in Eq. (11.131), we arrive at

$$\cos \varsigma = \pm \frac{\cos i \cos i' + \sqrt{\sin^2 i - \sin^2 \lambda}\sqrt{\sin^2 i' - \sin^2 \lambda}}{1 - \sin^2 \lambda}. \quad (11.136)$$

Alternatively, in terms of cosines only,

$$\cos \varsigma = \pm \frac{\cos i \cos i' + \sqrt{\cos^2 \lambda - \cos^2 i}\sqrt{\cos^2 \lambda - \cos^2 i'}}{\cos^2 \lambda}. \quad (11.137)$$

In the above equations, the + sign corresponds to $i' > i$ whereas the — sign corresponds to $i' < i$ on northbound orbits with the opposite sense in the southbound orbits [cf. Badhwar *et al.* (1990)].

The true anomaly θ' of the fragment, which dictates the sign of dv_r, is determined from the argument of latitude u' and the argument of perigee ω' of the fragment at the time of the breakup as

$$\theta' = u' - \omega'. \quad (11.138)$$

Applying the law of sines to the spherical triangle $A'BC$ of Fig. 11.3, one obtains

$$u' = \sin^{-1} \frac{\sin \lambda}{\sin i'}. \quad (11.139)$$

One should note that Eq. (11.139) corresponds to the northbound motion of the fragment at the time of breakup. For southbound motion, one needs to take the complement of u' from π. Since the argument of perigee is perturbed by the oblateness of the Earth, ω' at the time of observation may already have deviated from that at the time of fragmentation. Equation (11.107) provides the necessary correction.

As alternatives to Eqs. (11.123), (11.127) and (11.128), the velocity perturbations components can be obtained from Lagrange's planetary equations (11.79)–(11.81). Between Eqs. (11.79) and

(11.80), one obtains by elimination:

$$dv_r = \frac{na^2\sqrt{1-e^2}}{2er^2\sin\theta}\left[2aede - \left(1 - e^2 - \frac{r^2}{a^2}\right)da\right], \qquad (11.140)$$

and

$$dv_\theta = \frac{na\sqrt{1-e^2}}{2r}\left[da - \frac{2ae}{1-e^2}de\right]. \qquad (11.141)$$

From Eq. (11.81), one obtains by inversion:

$$dv_h = \frac{na^2\sqrt{1-e^2}}{r\cos u}di. \qquad (11.142)$$

Equations (11.140)–(11.142) were widely-used in orbital debris studies [cf. Johnson and McKnight (1987)]. However, their shortcomings were demonstrated by Tan (1987). The exact solutions (11.123), (11.127) and (11.128) were subsequently obtained by R. C. Reynolds [cf. Badhwar *et al.* (1990)].

Exercises

11.1. Derive Eqs. (11.7) and (11.8).

11.2. Derive or verify all equations of Sec. 11.3.

11.3. Derive or verify all equations of Sec. 11.4.

11.4. Derive or verify all equations of Sec. 11.5.

11.5. Derive or verify all equations of Sec. 11.6.

11.6. Derive or verify all equations of Sec. 11.7.

11.7. Derive or verify all equations of Sec. 11.8.

11.8. Verify Eqs. (11.79) to (11.84).

11.9. Derive or verify all equations of Sec. 11.11.

11.10. Obtain the semi-major axis of a Molniya orbit. Given a typical apogee height of 39,500 km, find the perigee height. A Molniya satellite gives about 8 h of coverage each of Russia and North America in a period of 24 h [cf. Verger *et al.* (2003)].

11.11. Express the number of revolutions per day of a satellite N in a Molniya orbit in terms of the solar day P (24 h) and the sidereal rotation period of the Earth P_s (23.9345 h).

11.12. Derive or verify all equations of Sec. 11.14.

11.13. Discuss the singularitites in Eqs. (11.140) and (11.142), which render them unreliable (Tan, 1987).

Appendix

The Ellipse and
its Properties

A.1 The Ellipse in Everyday Life

The *ellipse* is a natural shape which is encountered in everyday life. Any circular outline, when viewed at an angle appears as an ellipse. The basketball hoop, for example, appears elliptical to a player, unless, of course, when the player is vertically below the hoop. Another example arises when a drink in a paper cup is tilted from the vertical — the horizontal outline of the liquid takes an elliptical shape [cf. Fig. A.1, from Ogilvy (1956)].

The elliptical shape can be traced out on a paper in the following demonstrations. In the first one, torchlight is shone at an angle on a piece of paper on a table — the outline of the bright area is an ellipse. This is because the torchlight in three-dimensional space is conical in shape, and its intersection with the plane of the paper produces an elliptical outline (Fig. A.2). Alternatively, a spherical ball may be placed on the paper. Light from a single electric bulb at an angle will cast an elliptical shadow on the paper below [Fig. A.3 from Ogilvy (1956)]. This is again due to the fact that the spherical ball casts a shadow in space which is conical in shape. In this case, the point of

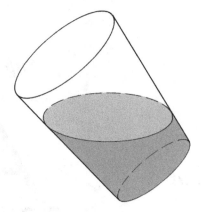

Fig. A.1

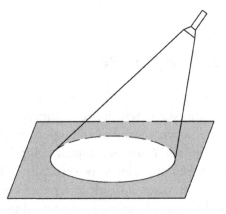

Fig. A.2

contact of the ball with the paper marks the location of one focus of the ellipse [cf. Ogilvy (1956)].

The ellipse is widely viewed as the section of right circular cone by a plane at an angle. Ellipses are also obtained as the intersection of a circular cylinder with a plane [cf. Tan (2001)]. In the earlier example of the liquid in a tilted paper cup, its outline would still be elliptical if the paper cup were a part of a circular cylinder. In our solar system, the shadow cast by Saturn on its rings (when its axis is tilted towards or away from the Sun) is nearly elliptical since

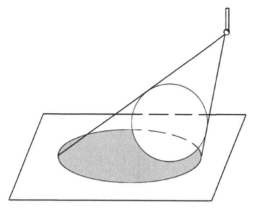

Fig. A.3

the shadow cast by Saturn in space is nearly circular cylindrical. However, because of the fact that Saturn is significantly oblate in shape, there is a very miniscule deviation of the shadow from the circular cylindrical shape and hence the consequent deviation of the shadow on the rings from the elliptical shape.

A.2 The Definition of an Ellipse

Definition. The ellipse is defined as the locus of a point which moves such that the ratio of its distance from a fixed point to that from a fixed line is a constant of less than one. The fixed point is called the *focus*; and the fixed line is called the *directrix*. The ellipse lies on a plane defined by the focus and the directrix. In Fig. A.4, P is any general point point on the ellipse, F is the focus, and D is the directrix. By the definition,

$$\frac{\overline{PF}}{\overline{PD}} = e, \tag{A.1}$$

where $0 \le e < 1$. e is called the *eccentricity* of the ellipse.

The resulting curve looks like an *oval* or an oblong circle. It has a *center* C, a largest diameter $V'V$ called the *major axis*, and

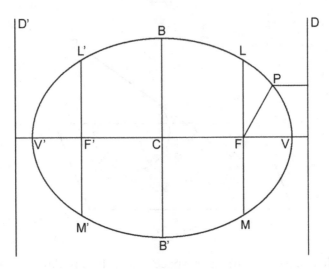

Fig. A.4

a smallest diameter perpendicular to the major axis BB' called the *minor axis* (Fig. A.4). (A diameter is defined as a chord passing through the center). The ellipse is symmetrical about the major and minor axes. There is thus a second focus F' and a second directrix D' (Fig. A.4). The two points of the ellipse on the major axis (V and V') are called the **vertices**. The two chords passing through the foci and perpendicular to the major axis (LM and $L'M'$) are called the latera recta (plural of *latus rectum*).

It should be noted that the same definition applies to the other conic sections having different eccentricities, viz., the *parabola* ($e = 1$) and the *hyperbola* ($e > 1$). The *circle* ($e = 0$) is considered a limiting case of the ellipse. Only the circle and the ellipse are closed curves which permit stable planetary orbits as opposed to the parabola and the hyperbola, which give open trajectories. The circle and the parabola are similar curves, i.e., any two circles (or parabolas, for that matter) have the same shape. Thus any two circles can be made to coincide by a proper magnification and moving their centers to the same location. Likewise, any two parabolas can be made to coincide by a proper translation, rotation and magnification. Ellipses (and hyperbolas) are not similar. However, ellipses (or hyperbolas) having the same eccentricities are similar.

A.3 Special Properties of the Ellipse

The ellipse has some unique properties, some of which are given as follows.

Property 1. The sum of the focal distances of any point on the ellipse is constant and equal to the major axis. Referring to Fig. A.5,

$$\overline{PF} + \overline{PF'} = 2a, \tag{A.2}$$

where $2a$ is the major axis and a is the **semi-major axis**. This is a fundamental property of the ellipse, which can be used as an alternative definition of the ellipse. The ellipse is the locus of a point which moves in a plane such that the sum of its distances from two fixed points on that plane is a constant. Note that this property is unique to the ellipse and does not apply to the other

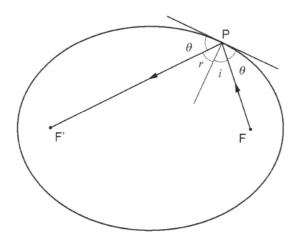

Fig. A.5

conic sections. Consequently, the alternative defintition furnished by Eq. (A.2), unlike Eq. (A.1), is for ellipses only.

Property 2. The tangent to the ellipse at any point makes equal angles with the focal radii at that point. In Fig. A.5, the two angles θ are equal. This is also referred to as the **reflective property** of the ellipse. Any ray of light passing through a focus F inside an elliptical reflector will emerge through the other focus F' following the law of reflection: $i = r$. Thus, a light source located at one focus will form its image at the other focus inside an elliptical reflector.

Property 3. The perpendicular from a focus upon the tangent at any point on the ellipse will intersect the corresponding directrix at the same point as the radial line of the point from the center (Fig. A.6).

Property 4. The perpendiculars from the foci on any tangent will intersect the tangent at points on the **auxiliary circle**, which circumscribes the ellipse (Fig. A.6).

Property 5. The product of the focal perpendiculars on the tangent at any point is constant and equal to the square of the semi-minor

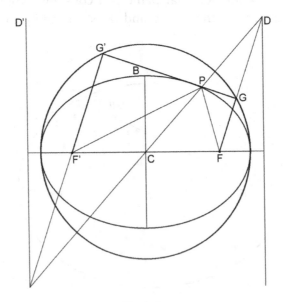

Fig. A.6

axis. Referring to Fig. A.6,

$$\overline{FG}.\overline{F'G'} = \overline{CB}^2. \tag{A.3}$$

A.4 An Ellipse as Sections of Cones and Cylinders

The circle, ellipse, parabola and hyperbola are known as the **conic sections** and are obtained as intersections of a plane and a right circular cone at various angles. If the intersection is a closed outline, then it is an ellipse. This includes the circle, which is considered a special case of an ellipse. The locations of the foci of the ellipse are found by inscribing two spheres which touch the cone and the plane of the ellipse on either side of it [cf. Fig. A.7, from Ogilvy (1956)].

An ellipse is also obtained as an intersection of a plane and a right circular cylinder at an angle (Fig. A.8). In this case also, the locations

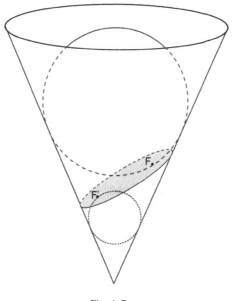

Fig. A.7

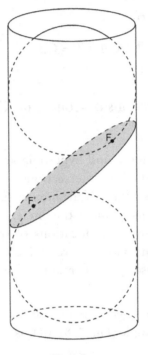

Fig. A.8

of the foci are found by inscribing two spheres in the cylinder, touching the plane on either side [cf. Tan (2001)]. This method has the added advantage that an ellipse with a pre-determined eccentricity e can be obtained by slicing the cylinder at an angle of $\pi/2 - \sin^{-1}\theta$ from the axis of the cylinder (Tan, 2001). The parabola and the hyperbola cannot be obtained in this manner.

A.5 Constructions of the Ellipse

There are several methods of constructing an ellipse. A few are outlined below.

Method 1. The ellipse can be thought of as a flattened circle, the degree of flattening depending upon the eccentricity of the ellipse.

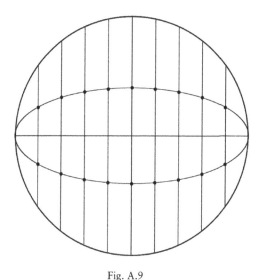

Fig. A.9

For a desired ellipse of eccentricity e, the minor axis bears to the major axis a ratio of $r = \sqrt{1 - e^2}$. Draw the auxiliary circle and a diameter equal to the major axis of the ellipse (Fig. A.9). Draw chords perpendicular to the diameter and shrink the chords by the ratio r. Then the locus of the ends of the chords give the desired ellipse.

Method 2. This method, based on the special property 1 above, is called the ***string and pins construction***. For a desired ellipse having semi-major axis a and eccentricity e, take a string of length $2a$ and attach the two ends to two pins fixed on a piece of paper, a distance $2ae$ apart. With a pencil, trace out a curve, keeping the string taut (Fig. A.10). The curve will be the desired ellipse with the pins giving the locations of the foci. Alternatively, one can take a loop of string of length $2a + 2ae$ and wind it around the two pins and use a pencil to trace out the curve. The result will be the same.

Method 3. For a desired ellipse having a semi-major axis a and semi-minor axis $b = a\sqrt{1 - e^2}$, a ***straight-edge construction*** is used. Take a straight edge AB of length $a + b$, and mark a point P, dividing the straight edge into parts having lengths equal to a and b. Place one

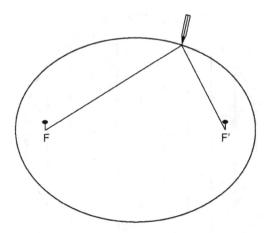

Fig. A.10

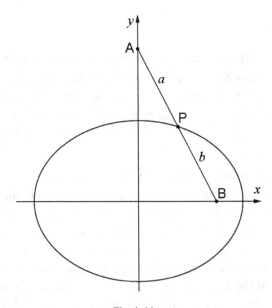

Fig. A.11

end of the straight edge (A) on the y-axis and the other end (B) on the x-axis and mark the location of P on the paper (Fig. A.11). Move the straight edge to trace out the locus of the point P. Repeat for all quadrants to complete the ellipse.

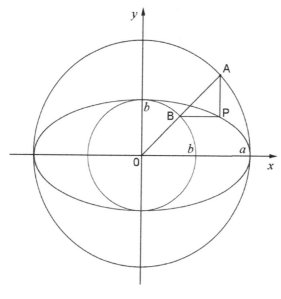

Fig. A.12

Method 4. In this method, two concentric circles having radii equal to the semi-major and semi-minor axes, a and b, are drawn (Fig. A.12). The larger circle will circumscribe the project ellipse and the smaller circle will be inscribed in it. Draw a radial line OA to intersect the two circles at A and B, respectively. Drop a vertical line AP from A and draw a horizontal line BP from B. The two lines intersect on a point P on the ellipse. Repeat the procedure to trace out the entire ellipse. This method has been termed a ***two-circle construction*** [cf. Gellert *et al.* (1977)].

Method 5. To obtain an ellipse having semi-major axis a and eccentricity e, draw the auxiliary circle, a diameter to be the major axis, and locate the two foci, at distances of ae on either side of the center. Draw parallel lines FA and $F'B$ to meet the circle at A and B, respectively (Fig. A.13). Repeat this procedure many times for different points on the circle. The envelope of AB gives the desired ellipse. This method is based on the special property 4 of the ellipse and is called the ***Rytz construction*** [cf. Gellert *et al.* (1977)].

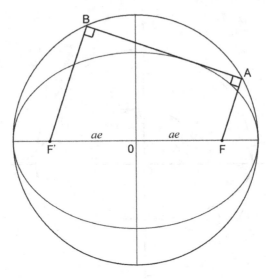

Fig. A.13

Method 6. For an ellipse having semi-major axis a and eccentricity e, draw a circle of radius $2a$, with F as center and CD as a diameter; choose a point F' at a distance of $x = 2ae$ from F; draw several lines from F' to the circle; and draw perpendicular bisectors of these lines (Fig. A.14). The envelope of the bisectors will give the desired ellipse,

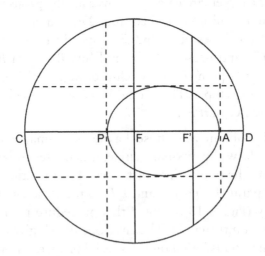

Fig. A.14

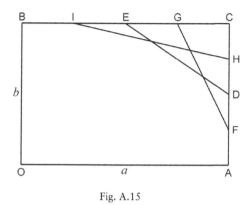

Fig. A.15

with the focal points at F and F'. In Fig. A.14, only the perpendicular bisectors parallel with and perpendicular to the x- and y-axes are shown. It is easy to verify that

$$\overline{CP} = a(1 + e), \tag{A.4}$$

and

$$\overline{PF} = \overline{F'A} = \overline{AD} = a(1 - e). \tag{A.5}$$

Method 7. One can obtain an ellipse having semi-major axis a and semi-minor axis b in quarters, as follows. Draw a rectangle $OACB$ with $\overline{OA} = a$ and $\overline{OB} = b$ (Fig. A.15). Join the mid-points of AC and BC, which are D and E, respectively. Next, join the mid-points of AD and EC (F and G) and those of DC and BE (H and I) (Fig. A.14). Continue in this fashion. The envelope of the lines will rapidly outline one quarter of the desired ellipse. The remaining three quarters can be similarly constructed.

A.6 General Equation of a Conic in Cartesian Coordinates

The general equation of a conic in Cartesian coordinates (x, y) is an equation of the second degree:

$$Ax^2 + Bxy + Cy^2 + Dx + Ey + F = 0. \tag{A.6}$$

The nature of the conic depends on the **discriminant**,

$$\Delta = \begin{vmatrix} 2A & B & D \\ B & 2C & E \\ D & E & 2F \end{vmatrix}, \tag{A.7}$$

and the quantity,

$$\delta = B^2 - 4AC. \tag{A.8}$$

The conic degenerates into a pair of parallel or intersecting straight lines if $\Delta = 0$. The conic is proper if and only if $\Delta \neq 0$. In the latter case, the conic is a parabola if $\delta = 0$; an ellipse if $\delta < 0$; and a hyperbola if $\delta > 0$.

A.7 Equations of an Ellipse in Cartesian Coordinates

When the axes of the ellipse are parallel with the coordinate axes, $B = 0$. Then the general equation of the ellipse reduces to

$$Ax^2 + Cy^2 + Dx + Ey + F = 0. \tag{A.9}$$

If further, the center of the ellipse is at the origin with the major and minor axes coinciding with the x- and y-axes, respectively, then one obtains the **standard form of the equation of ellipse** in Cartesian coordinates [cf. Oakley (1957)]:

$$\frac{x^2}{a^2} + \frac{y^2}{b^2} = 1. \tag{A.10}$$

Here a is the semi-major axis and b is the semi-minor axis of the ellipse.

Equation (A.10) can be obtained from the alternative definition of the ellipse given by Eq. (A.2). Denoting the coordinates of the foci

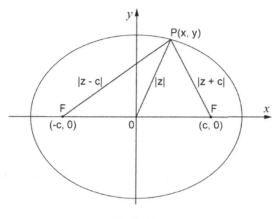

Fig. A.16

F and F' as $(c, 0)$ and $(-c, 0)$, respectively (Fig. A.16), we have

$$\sqrt{(x + c)^2 + y^2} + \sqrt{(x - c)^2 + y^2} = 2a. \qquad \text{(A.11)}$$

The standard technique consists of the following steps [cf. Weisstein (2003)]: firstly, move the second term to the right and square both sides to obtain

$$(x + c)^2 + y^2 = 4a^2 - 4a\sqrt{(x - c)^2 + y^2} + (x - c)^2 + y^2. \quad \text{(A.12)}$$

Secondly, solve for the square root term and simplify to obtain

$$\sqrt{(x - c)^2 + y^2} = a - \frac{cx}{a}. \qquad \text{(A.13)}$$

Thirdly, square once more and simplify to obtain

$$\frac{x^2}{a^2} + \frac{y^2}{a^2 - c^2} = 1. \qquad \text{(A.14)}$$

By setting $b^2 = a^2 - c^2$, we get Eq. (A.10).

In terms of a and e, Eq. (A.8) takes the form

$$x^2(1 - e^2) + y^2 = a^2. \tag{A.15}$$

The equation of the tangent to a point (x', y') on the ellipse is given by [cf. Geometry Problem Solver (1977)]

$$\frac{xx'}{a^2} + \frac{yy'}{b^2} = 1. \tag{A.16}$$

Differentiating Eq. (A.16) with respect to x, we obtain the slope of the tangent to the ellipse at (x', y'):

$$\frac{dy}{dx} = -\frac{b^2 x'}{a^2 y'}. \tag{A.17}$$

For example, at the top end of the latus rectum through the right focus, we have $x' = ae$ and $y' = a(1 - e^2)$. Thus,

$$\frac{dy}{dx} = -e, \tag{A.18}$$

a result obtained in Sec. 1.5.

A.8 Equation of an Ellipse in Cartesian Coordinates with the Right Focus at the Origin

If the origin of the coordinate system is moved to the right focus of the ellipse (Fig. A.17), the equation of the ellipse takes the form [cf. Sec. 6.2]

$$\sqrt{x^2 + y^2} + ex = p. \tag{A.19}$$

In the (a, e) representation, this is written as

$$\sqrt{x^2 + y^2} + ex = a(1 - e^2). \tag{A.20}$$

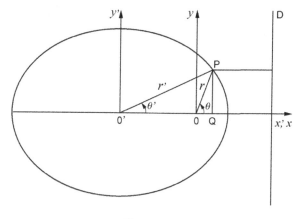

Fig. A.17

The slope of the tangent at any point on the ellipse is obtained by differentiating Eq. (A.19) with respect to x:

$$\frac{dy}{dx} = -\frac{e\sqrt{x^2 + y^2}}{y} - \frac{x}{y}. \tag{A.21}$$

We get the result of Eq. (A.18) by setting $x = 0$.

A.9 Parametric Equations of the Ellipse

It is often advantageous to express x and y in terms of a single variable, called the **parameter**. In the case of the ellipse, we can write it in terms of the parameter θ,

$$x = a\cos\theta, \tag{A.22}$$

and

$$y = b\sin\theta. \tag{A.23}$$

The standard equation of the ellipse (A.10) readily follows from Eqs. (A.22) and (A.23).

Another form of parametric equations is found in the variable t defined by [cf. Hartley (1960)]

$$\tan \frac{\theta}{2} = t. \tag{A.24}$$

Then,

$$\sin \theta = \frac{2t}{1+t^2}, \tag{A.25}$$

and

$$\cos \theta = \frac{1-t^2}{1+t^2}. \tag{A.26}$$

The parametric equations (A.22) and (A.23) assume the forms

$$x = a\frac{1-t^2}{1+t^2}, \tag{A.27}$$

and

$$y = b\frac{2t}{1+t^2}. \tag{A.28}$$

A.10 The Polar Equation of a Conic

The general equation of a conic in polar coordinates (r, θ) is written as

$$r = \frac{p}{1+e\cos\theta}. \tag{A.29}$$

This form has the great advantage that the same equation fits all conics, depending upon the eccentricity according to the following scheme: circle, $e = 0$; ellipse, $0 \le e < 0$; parabola, $e = 1$; and hyperbola, $e > 1$. In the case of the ellipse, p represents the semi-latus rectum. The right focus of the ellipse is located at the origin (Fig. A.17).

One can arrive at Eq. (A.29) from the general definition of a conic in Sec. A.2. In Fig. A.16, let the distance between the focus O and the directrix D be k. Then from Eq. (A.1),

$$\overline{OP} = e \cdot \overline{PD} = e(\overline{OD} - \overline{OQ}), \tag{A.30}$$

or

$$r = e(k - r\cos\theta). \tag{A.31}$$

Rearranging and letting $ek = p$, we obtain Eq. (A.29).

In terms of a and e, the polar equation is alternatively written in the form

$$r = \frac{a(1 - e^2)}{1 + e\cos\theta}. \tag{A.32}$$

A.11 The Pedal Equation of the Ellipse

The pedal equation of the ellipse follows from Properties 1 and 5 of the ellipse in Sec. A.3 [cf. Hartley (1960)]. In Fig. A.6, we have, from the similar triangles FPG and $F'PG'$:

$$\frac{\overline{FG}}{\overline{FP}} = \frac{\overline{F'G'}}{\overline{F'P}}. \tag{A.33}$$

Expressing the pedal distances $\overline{FG}$ and $\overline{F'G'}$ by ρ and ρ' respectively, $\overline{CB}$ by b and $\overline{FP}$ by r, one obtains the pedal equation of the ellipse, from Eqs. (A.2), (A.3) and (A.33):

$$\frac{b^2}{\rho^2} = \frac{2a}{r} - 1. \tag{A.34}$$

A.12 The Equation of an Ellipse in Eccentric Polar Coordinates

If the center of the ellipse is taken as the origin of coordinates, then we have what is quite inappropriately termed as the eccentric polar coordinate (r', θ'), where θ' is called the *eccentric angle* [cf. Weisstein (2003)]. We have, from Fig. A.17,

$$x' = r' \cos \theta', \tag{A.35}$$

and

$$y' = r' \sin \theta'. \tag{A.36}$$

In eccentric polar coordinates, the equation of the ellipse takes the form [cf. Weisstein (2003)]

$$r' = a \sqrt{\frac{1 - e^2}{1 - e^2 \cos^2 \theta'}}. \tag{A.37}$$

A.13 The Equation of an Ellipse in Complex Coordinates

The equation of an ellipse, with its center at the origin and major axis on the abscissa, can be written in terms of complex coordinates (vide Fig. A.16) as [cf. Speigel (1964)]

$$|z + c| + |z - c| = 2a. \tag{A.38}$$

In polar form, the equation

$$z = ae^{i\theta} + be^{-i\theta}, \tag{A.39}$$

represents an ellipse with the major axis vertices located at $z = \pm(a + b)$ and minor axis vetices located at $z = \pm i(a-b)$ with

$0 \leq \theta < 2\pi$. An ellipse having a semi-major axis a and semi-minor axis b is described by the following equation [cf. Price (2002)]:

$$z = \frac{a+b}{2}e^{i\theta} + \frac{a-b}{2}e^{-i\theta}. \qquad (A.40)$$

The vertices are located at $z = \pm a$; and $z = \pm ib$.

References

Altman, S.P., *Orbital Hodograph Analysis*, AAA Science and Technology Series, Western Periodicals, N. Hollywood, 1965.

Aravind, P.K., *Am. J. Phys.*, *55*, 1144–1145, 1987.

Arfken, G.B. and Weber, H.J., *Mathematical Methods for Physicists*, Elsevier, Amsterdam, 2005.

Badhwar, G.D., Tan, A. and Reynolds, R.C., *J. Spacecr. Rockets*, *27*, 299–305, 1990.

Bartlett, A.A. and Concklin, R.L., *Am. J. Phys.*, *53*, 242–248, 1985.

Bate, R.R., Mueller, D.D. and White, J.E., *Fundamentals of Astrodynamics*, Dover, New York, 1971.

Born, M., *Atomic Physics*, Blackie and Son, London, 1969.

Boyd, J.N. and Raychowdhury, P.N., *Am. J. Phys.*, *53*, 502, 1985.

Brouwer, D. and Clemence, G.M., *Methods of Celestial Mechanics*, Academic Press, New York, 1961.

Bucher, M. and Siemens, D.P., *Am. J. Phys.*, *66*, 88–89, 1998.

Chenciner, A. and Montgomery, R., *Ann. Math.*, *152*, 881–901, 2000.

Cohen, I.B., *Sci. Amer.*, *244*, 166–179, 1981.

Cook, A.H., *Interiors of the Planets*, Cambridge, 1980.

COSPAR International Reference Atmosphere, Akademie, 1972.

Courant, R. and Friederichs, K.O., *Superesonic Flow and Shock Waves*, Interscience Publishers, London, 1948.

Danby, J.M.A., *Fundamentals of Celestial Mechanics*, Willmann-Bell, Richmond, 1988.

Dermott, S.F., *Nature Phys. Sci.*, 244, 18–21, 1973.

Eddington, A.S., *The Mathematical Theory of Relativity*, Cambridge, 1963.

Ehricke, K.A., *Space Flight II. Dynamics*, Van Nostrand, Princeton, 1962.

Finkel, R.W., *Am. J. Phys.*, 58, 1085–1087, 1990.

Finlay-Freundlich, E., *Celestial Mechanics*, Pergamon Press, New York, 1958.

Fowles, G.R., *Analytical Mechanics*, Holt, Reinhart and Winston, New York, 1962.

Freeman, I., *Am. J. Phys.*, 45, 585–586, 1977.

Gellert, W., Kuestner, H., Hellwich, M. and Kaestner, H., *The VNR Encyclopedia of Mathematics*, Van Nostrand, New York, 1977.

Geometry Problem Solver, Research and Education Association, New York, 1977.

Goldstein, H., *Classical Mechanics*, Addison-Wesley, Reading, 1980.

Greenwood, D.T., *Classical Mechanics*, Dover, New York, 1977.

Hartley, E.M., *Cartesian Geometry of the Plane*, Cambridge, 1960.

Hedin, A.E., *J. Geophys. Res.* 88, 1983.

Iles, K. and Wilson, L.J., *Math. Teach.*, 73, 32–34, 1980.

Jacchia, L.G., *Smithsonian Astrophys. Obs. Spec. Rept.* 332, 1971.

Jacchia, L.G., *Smithsonian Astrophys. Obs. Spec. Rept.* 375, 1977.

Johnson, N.L., *J. Brit. Interplanet. Soc.*, 36, 51–58, 1983a.

Johnson, N.L., *J. Brit. Interplanet. Soc.*, 36, 357–362, 1983b.

Johnson, N.L., *The Orbital Debris Quarterly News*, 1(2), 1, 1996.

Johnson, N.L. and McKnight, D.S., *Artificial Space Debris*, Orbit Book Co., Malabar, 1987.

Johnson, N.L., Whitlock, D.O., Anz-Meador, P., Cizek, M.E. and Portman, S.A., *History of On-Orbit Satellit Fragmentations*, NASA Johnson Space Center, 2004.

King-Hele, D., *Theory of Satellite Orbits in an Atmosphere*, Butterworths, London, 1964.

King-Hele, D., *Satellite Orbits in an Atmosphere: Theory and Applications*, Blackie, Glasgow, 1987.

Lamb, H., *Dynamics*, Cambridge, 1961.

Lawrence, J.P., *A Catalog of Special Plane Curves*, Dover, New York, 1972.

Lim, Y.-K. (ed.), *Problems and Solutions on Mechanics*, World Scientific, Singapore, 1994.

Lockwood, E.H., *A Book of Curves*, Cambridge, 1961.

Macklin, P.A., *Am. J. Phys.*, 39, 1088–1089, 1971.

MacMillan, W.D., *Statics and Dynamics of a Particle*, McGraw-Hill, New York, 1927.

Marion, J.B. and Thornton, S.T., *Classical Dynamics of Particles and Systems*, Saunders, Fort Worth, 1995.

Mathematica, Wolfram Research, 1997.

McBride, N. and Gilmour, I., *An Introduction to the Solar System*, Cambridge, 2003.

Meirovitch, L., *Methods of Analytical Dynamics*, McGraw-Hill, New York, 1970.

Moore, C., *Phys. Rev. Lett.*, 70, 3675–3679, 1993.

Morse, P.M. and Feshbach, H., *Methods of Theoretical Physics*, McGraw-Hill, New York, 1953.

Moulton, F.R., *An Introduction to Celestial Mechanics*, Dover, New York, 1970.

Murray, C.D. and Dermott, S.F., *Solar System Dynamics*, Cambridge, 1999.

Nelson, W.C. and Loft, E.E., *Space Mechanics*, Prentice-Hall, Englewood Cliffs, 1962.

Norwood, J., *Intermediate Classical Mechanics*, Prentice-Hall, Englewood Cliffs, 1979.

Ogilvy, C.S., *Excursions in Mathematics*, Dover, New York, 1956.

Osgood, W.F., *Mechanics*, Dover, New York, 1965.

Price, T.E., *Math. Mag.*, 75, 300–307, 2002.

Prussing, J.E., *Am. J. Phys.*, 45, 1216–1217, 1977.

Salmon, G., *A Treatise on Conic Sections*, Chelsea, New York, 1954.

Sandin, T.R., *Phys. Teach.*, 28, 36–38 (1990).

Spiegel, M.R., *Theory and Principles of Complex Variables*, McGraw-Hill, New York, 1964.

Stein, S.K., *Math. Mag.*, 50, 160–162, 1977.

Tan, A., *Am. J. Phys.*, 47, 741–742, 1979a.

Tan, A., *Eureka*, 40, 30–32, 1979b.

Tan, A., *Eureka*, 41, 17–22, 1981.

Tan, A., *Math. Teach.*, 75, 638, 666, 1982.

Tan, A., in *NASA CR 172009*, W.B. Jones and S.H. Goldstein (ed.), 1987.

Tan, A., *Am. J. Phys.*, 56, 501, 588, 1988.

Tan, A., *Theta*, 5(1), 7–9, 1991.

Tan, A., *Theta*, 6(1), 27–31, 1992a.

Tan, A., *Theta*, 6(2), 15–19, 1992b.

Tan, A., *Theta*, 8(1), 11–16, 1994.

Tan, A., in *NASA CR 188410*, R. Bannerot and D.G. Sickorez (ed.), 1995.

Tan, A., *Math. Gazette*, 85, 491–492, 2001.

Tan, A., *Math. Educ.*, 38, 63–70, 2004.

Tan, A., *Math. Spectrum*, 37, 126–130, 2004–2005.

Tan, A. and Chameides, W.L., *Am. J. Phys.*, 49, 691–692, 1981.

Tan, A. and Ramachandran, R., *J. Astronaut. Sci.*, 53, 39–50, 2005.

Tan, A. and Zhang, D., *J. Astronaut. Sci.*, 49, 585–599, 2001.

Tan, A., Badhwar, G.D., Allahdadi, F.A. and Medina, D., *J. Spacecr. Rockets, 33*, 79–85, 1996.

The New Solar System, J.K. Beatty, C.C. Peterson, and A. Chaikin, Cambridge, 1999.

The Orbital Debris Quarterly News, 7(3), 1, 2002.

Timoshenko, S. and Young, D.H., *Advanced Dynamics*, McGraw-Hill, New York, 1948.

Van de Kamp, P., *Elements of Astromechanics*, Freeman, San Francisco, 1964.

Van de Kamp, P., *Principles of Astrometry*, Freeman, San Francisco, 1967.

Verger, F., Sourbas-Verger, I. and Ghirardi, R., *The Cambridge Encyclopedia of Space*, Cambridge, 2003.

Weinstock, R., *Am. J. Phys.*, *30*, 813–814, 1962.

Weinstock, R., *Am. J. Phys.*, *40*, 357–358, 1972.

Weinstock, R., *Am. J. Phys.*, *60*, 615–619, 1992.

Weisstein, E.W., *CRC Concise Encyclopedia of Mathematics*, Chapman and Hall, Boca Raton, 2003.

Whittaker, E.T., *A Treatise on the Analytical Dynamics of Particles and Rigid Bodies*, Cambridge, 1961.

Index